Computational Models for Neuroscience

Springer
London
Berlin
Heidelberg
New York
Hong Kong
Milan
Paris
Tokyo

Robert Hecht-Nielsen and Thomas McKenna (Eds)

Computational Models for Neuroscience

Human Cortical Information Processing

Springer

BS

Robert Hecht-Nielsen, PhD
University of California, San Diego, CA, USA
HNC Software Inc, San Diego, CA, USA

Thomas McKenna, PhD
Office of Naval Research, Arlington, VA, USA

British Library Cataloguing in Publication Data
Computational models for neuroscience : human cortical
   information processing
   1. Cerebral cortex - Computer simulation 2. Computational
   Neuroscience
   I. Hecht-Nielsen, Robert II. McKenna, Thomas M.
   006.3
ISBN 1852335939

Library of Congress Cataloging-in-Publication Data
Computational models for neuroscience / Robert Hecht-Nielsen and Thomas
McKenna (eds).
   p. cm.
   Includes bibliographical references and index.
   ISBN 1-85233-593-9 (alk. Paper)
   1. Cerebral cortex - - Computer simulation. 2. Computational neuroscience. 3. Neural
   networks (Neurobiology) 4. Neural networks (Computer science) I. Hecht-Nielsen,
   Robert. II. McKenna, Thomas.
   QP383.C65 2002
   612.8'25-dc21

                                                            2002070455

ISBN 1-85233-593-9 Springer-Verlag London Berlin Heidelberg
a member of BertelsmannSpringer Science+Business Media GmbH
http://www.springer.co.uk

Typesetting: Electronic text files prepared the editors
Printed and bound at the Athenæum Press Ltd., Gateshead, Tyne & Wear
34/3830-543210 Printed on acid-free paper SPIN 10867381

6 |12|06

Preface

Formal study of neuroscience (broadly defined) has been underway for millennia. For example, writing 2,350 years ago, Aristotle[1] asserted that association – of which he defined three specific varieties – lies at the center of human cognition. Over the past two centuries, the simultaneous rapid advancements of technology and (consequently) per capita economic output have fueled an exponentially increasing effort in neuroscience research.

Today, thanks to the accumulated efforts of hundreds of thousands of scientists, we possess an enormous body of knowledge about the mind and brain. Unfortunately, much of this knowledge is in the form of isolated factoids. In terms of "big picture" understanding, surprisingly little progress has been made since Aristotle. In some arenas we have probably suffered negative progress because certain neuroscience and neurophilosophy precepts have clouded our self-knowledge; causing us to become largely oblivious to some of the most profound and fundamental aspects of our nature (such as the highly distinctive propensity of all higher mammals to automatically segment all aspects of the world into distinct holistic objects and the massive reorganization of large portions of our brains that ensues when we encounter completely new environments and life situations).

At this epoch, neuroscience is like a huge collection of small, jagged, jigsaw puzzle pieces piled in a mound in a large warehouse (with neuroscientists going in and tossing more pieces onto the mound every month). Few attempts to find pieces in the mound that fit together have succeeded. Based upon this record of failure, few such attempts are now launched. Neuroscience research today is largely centered on the activity of uncovering new factoids in a careful, systematic fashion and then tossing them on the mound.

---

[1] Aristotle (350 BC, 1995) Collected Works, Sixth Printing with Corrections, Princeton NJ: Princeton University Press.

Among neuroscientists there is a general anticipation that someday (not soon) somebody will fit enough pieces of the puzzle together to discern a key concept that will trigger rapid and sustained progress in assembling the rest of the existing pieces and in discovering the still missing pieces. The purpose of this book is to try to hurry up this timetable by describing several different conceptual frameworks that indicate how some of the pieces of the neuroscience puzzle may fit together. Perhaps the *trigger concept* that will lead to the rapid assembly of the neuroscience jigsaw puzzle is in this book. Or, if not, perhaps this book will inspire creation of the trigger concept.

Chapter 1 of this book, by Larry Cauller (the chapters are organized alphabetically by author name), presents a dynamics approach to understanding cortical function. As with many of the models and constructs presented here, Cauller's theory has been computer implemented and has been shown to have interesting and relevant properties. In particular, he concludes that it is the dynamical attractors of neuronal networks which are the fundamental, repeatable, basic elements of information storage and processing. Ideas such as neurons functioning as simple "feature detectors" are rejected. The net result is an iconoclast and fresh viewpoint about the workings of cerebral cortex.

Chapter 2, by Oleg Favorov and colleagues, takes a radically different viewpoint in comparison with Chapter 1; namely, that Frank Rosenblatt had it right all along. In this view, the brain is composed of perceptron-like multilayer neuron networks trained by backpropagation.[2] Favorov's theory is well argued. The main point is that this concept jibes with many known properties of cerebral information processing and behavior. The fact that "connectionist" cognitive science has been astoundingly successful in building phenomenological models of many cortical information processing operations is another strong argument for this theory.

Chapter 3, by Walter Freeman, is philosophically in tune with Chapter 1, but takes the main point even further, invoking chaos as a key ingredient in brain dynamics. In Freeman's conception, each neuron participates in multiple, interlinked, dynamical interactions with other neurons. Going farther than Cauller, Freeman's theory views trajectories transitioning from one attractor to another as the representations of information. The learning and behaving processes hinge on the occurrences of these transitions, not on the sustained attainment of stable attractor limit cycles. Appropriate (goal-satisfying) transitions are reinforced and enhanced, neutral or goal-denying transitions are suppressed. Freeman's viewpoint has heft because its main tenets are behaviors that real neurons can probably achieve and because it helps explain why conventional concepts have had such little success in explaining brain information processing.

Chapter 4, by Robert Hecht-Nielsen, presents an almost entirely new, detailed and complete, high-level theory of thalamocortical information processing. It postulates that all thalamocortical information processing can be accounted for by two particular types of associative memory neural networks, each employed in abundance. The

---

[2] Frank Rosenblatt invented the multilayer perceptron over forty years ago (Rosenblatt, F. (1962) Principles of Neurodynamics, New York: Spartan). However, he never found a way to adaptively train the hidden layer neurons. Backpropagation training of hidden layer weights was popularized by Rumelhart and his colleagues (Rumelhart, D.E., Hinton, G.E., and Williams, R.J. (1986) Learning representations by back-propagating errors, Nature 323: 533–536).

first of these networks is a fast-acting two-step attractor system (implemented by reciprocal connections between cortex and thalamus) which casts all information into one of a few thousand fixed-for-life sparse neuronal codes (*representational tokens*). The cerebral cortex is presumed to be tiled with tens of thousands of these (only weakly interacting) networks. The second network learns to connect certain pairs of these tokens on different regions. The strength of connection is a novel "fuzzy" information theoretic quantity called *antecedent support*; which Hecht-Nielsen claims is the only information learned, stored, and used in the thalamocortex. The notion that the cortex is fundamentally symbolic may appeal to old-school AI researchers. "Computational Intelligence" fans may like the theory because the learned relationships between pairs of hard symbols are soft and yet information-theoretic in character. Neuroscientists may like it because the whole thing seems to exactly match the known anatomy and physiology of the thalamocortex and its related structures.

Chapter 5, by Henry Markram, is a tour de force of advanced insight into those most critical brain elements: synapses. Markram's synthesis of the huge and confusing literature on this subject is guided by his own world-leading experimental findings. In particular, in strong resonance with several other chapters of the book, he emphasizes the central importance of dynamics in synaptic transmission: the post-synaptic effect of an arriving action potential depends critically on the recent history of arrivals of other action potentials. Going far beyond a mere recitation of the facts, Markram presents a compelling synthesis that identifies and characterizes principal categories of synaptic dynamical behavior. This chapter provides abundant knowledge of great value to all interested in cortical theory.

Chapter 6, by Thomas McKenna, provides sage commentary on a variety of cortical theories and their relationships. In particular, McKenna emphasizes and explores the gaps that clearly exist between cortical theories and biology. These gaps are then used as pointers towards potentially profitable new directions of theory construction. A valuable feature of McKenna's chapter is his characterizations and comparisons of theories that operate at different levels of abstraction. Going far beyond the simple "top down" and "bottom up" taxonomy, theories are discussed in terms of their ability to succeed at their individual level of abstraction. Two hundred years from now any high school student will probably be able to give a simple explanation for how the brain represents objects, learns and stores facts about those objects and then uses those facts to carry out information processing. A key point is that a correct theory can function without gaps at any level of detail. Another issue addressed in this chapter is the applicability of cortical theories. For theories to be judged correct and complete they must support practical implementations which can solve high-value human problems.

Chapter 7, by Jeffrey Sutton and Gary Strangman, proposes that the cerebral cortex can be viewed as being composed of hierarchical groups of neural networks; or, as they term it, a Network of Networks (NoN). The main point of this theory is that only modeling individual associative memory networks will never, by itself, explain the function of the cerebral cortex. A key missing ingredient is that groups of such networks are used, controlled, and trained as an ensemble – not individually. The chapter then goes on to explain how this might be accomplished and why this view is so important in understanding the obvious hierarchicality of processing we are so intimately familiar with (our cortex can deal with "throw the ball" as a conceptual unit just as easily as "slightly increase bicep tension").

Chapter 8, by John Taylor, presents a theory of cerebral information processing that concentrates on how consciousness may emerge from the combined activity of ensembles of stimulus-driven cortical-subcortical control structures set up to tend to a variety of behavioral goals and needs. Taylor's analysis frees itself from the details by fixating on firmly established high-level brain information properties that any successful theory must explain. Mixing together insights from a variety of theoretical perspectives, Taylor integrates these into a whole that emerges as a vision of how collections of relatively straightforward "thermostat" brain functions can, because of the strong causal linkages between their activating stimuli, end up supporting a higher-level process that becomes self-aware. Unlike past toy-system-level studies of "emergent" properties, Taylor's theory gets to the heart of why and how conscious awareness phylogenetically emerged. This chapter illustrates the value, and provides a masterful example, of theory building at a high conceptual level.

Chapter 9, by Neill Taylor and John Taylor, addresses another key cortex theory issue: language. In particular, they provide a foundational study of how a neuronal architecture can be developed to implement Noam Chomsky's hypothetical "Language Acquisition Device." Central to their theory are structures for chunking and unchunking of word strings (i.e., hierarchical networks). Using these, they develop a set of neuronally plausible language processing components and procedures that exhibit many of the key properties that a "LAD" must have. The resulting theory is not intended as a final theory of language acquisition and understanding. However, it provides a nice jumping-off point for further research.

Chapter 10, by Richard Zemel, is a theory of the population codes that many researchers believe lie at the heart of cortical information processing. In the past, almost all studies of population codes have assumed that the collection of neurons making up such a code are "ON" (and that other nearby neurons are "OFF"). Even with this binary assumption, many interesting properties of such codes have been established. In the theory presented in this chapter, Zemel goes way beyond this assumption: he allows population codes to also have variable non-negative activity levels assigned to them. He then shows that these activity levels can be interpreted in terms of a related probability distribution. Zemel then argues that the process of activating a population code is itself a probability calculation. Furthermore, he shows that collections of linked population codes can interact to form a type of belief network; a belief network that seems highly compatible with cortical neuronal anatomy, physiology, and function. The net result is a foundation for a new and promising direction of research.

The Editors thank all of the authors for their contributions.

Acknowledgements

The Editors would like to thank Kate Mark for compiling the book's chapters, completing the monumental job of uniformly formatting them, and carrying out the entire production process through to the printing of camera-ready copy. The early work of Myra Jose on the first draft of the book is also greatly appreciated. We thank Robert Means and Rion Snow for their assistance in indexing. Thanks also to Althera Allen for organizing and managing the workshop in La Jolla at which this book was conceived.

Robert Hecht-Nielsen
Thomas McKenna

August 2002

Acknowledgements

Contents

Chapter 6 The Development of Cortical Models to Enable Neural-based Cognitive Architectures 171

Chapter 7 The Behaving Human Neocortex as a Dynamic Network of Networks 205

Chapter 8 Towards Global Principles of Brain Processing... 221

Contributors

Chapter 1

Lawrence J. Cauller, PhD
    Neuroscience Department
    University of Texas, Dallas
    Richardson, Texas USA

Chapter 2

Oleg Favorov, PhD
    School of Electrical Engineering and Computer Science
    University of Central Florida
    Orlando, Florida USA

Dan Ryder
    Department of Philosophy
    University of North Carolina, Chapel Hill
    Chapel Hill, North Carolina USA

Joseph T. Hester
    Biomedical Engineering
    University of North Carolina, Chapel Hill
    Chapel Hill, North Carolina USA

Douglas G. Kelly
    Department of Statistics
    University of North Carolina, Chapel Hill
    Chapel Hill, North Carolina USA

Mark Tommerdahl
    Biomedical Engineering
    University of North Carolina, Chapel Hill
    Chapel Hill, North Carolina USA

Chapter 3

Walter J. Freeman
    Department of Molecular and Cell Biology
    University of California, Berkeley
    Berkeley, California USA

Chapter 4

Robert Hecht-Nielsen
    Department of Electrical and Computer Engineering
    and Institute for Neural Computation
    and Program in Computational Neurobiology
    University of California, San Diego
    La Jolla, California USA

Chapter 5

Henry Markram
    Department of Neurobiology
    Weizmann Institute for Science
    Rehovot, ISRAEL

Chapter 6

Thomas McKenna
    Office of Naval Research
    Arlington, Virginia USA

Chapter 7

Jeffrey P. Sutton, MD, PhD
    National Space Biomedical Research Institute and Neural Systems Group

Massachusetts General Hospital
Harvard-MIT Division of Health Sciences and Technology
Charlestown, Massachusetts USA

Gary Strangman, PhD
Massachusetts General Hospital Neural Systems Group
Harvard-MIT Division of Health Sciences and Technology
Charlestown, Massachusetts USA

Chapter 8

John G. Taylor
Department of Mathematics
Kings College
Strand, London UK

Chapter 9

Neill R. Taylor
Department of Mathematics
Kings College
Strand, London UK

John G. Taylor
Department of Mathematics
Kings College
Strand, London UK

Chapter 10

Richard Zemel
Department of Computer Science
University of Toronto
Toronto, Ontario CANADA

Chapter 1

The Neurointeractive Paradigm: Dynamical Mechanics and the Emergence of Higher Cortical Function

Larry Cauller

1.1 Abstract

Recently established biological principles of neural connectionism promote a neuro-interactivist paradigm of brain and behavior which emphasizes interactivity between neurons within cortical areas, between areas of the cerebral cortex, and between the cortex and the environment. This paradigm recognizes the closed architecture of the behaving organism with respect to motor/sensory integration within a dynamic environment where the majority of sensory activity is the direct consequence of self-oriented motor actions. The top-down cortical inputs to primary sensory areas, which generate a signal that predicts discrimination behavior in monkeys (Cauller and Kulics, 1991), selectively activate the cortico-bulbar neurons that mediate directed movements. Unlike the widely distributed axons and long-lasting excitatory synaptic effects of the top-down projections, which generate the associative context for motor/sensory interactivity, the bottom-up sensory projections are spatially precise and activate a brief excitation followed by a long-lasting inhibition (Cauller and Connors, 1994). Therefore, the sensory consequences of a motor action are the major source of negative feedback, which completes an interactive cycle of associative hypothesis testing: a winner-take-all motor/sensory pattern initiates a behavioral action within a top-down associative context; the bottom-up sensory consequences of that action interfere with top-down sensory predictions and strengthen or refine the associative hypothesis; then the testing cycle repeats as the sensory negative feedback inhibits the motor/sensory pattern and releases the next winner-take-all action.

Given this neurointeractivity, perception is a proactive behavior rather than information processing, so there is no need to impose representationalism: neurons simply respond to their inputs rather than encode sensory properties; neural activity patterns are self-organized dynamical attractors rather than sensory driven transformations; action is based upon a purely subjective model of the environment rather than a reconstruction. The associative hypothesis is the neurointeractive equivalent to awareness and hypothesis testing is the basis for attention. This neurointeractive process of action/prediction association explains early development: from self-organized cortical attractors in utero; to the emergence of self-identity in the newborn, who learns to predict the immediate effects of self-action (i.e. listening to its own speech sounds); to the discovery of ecological contingencies; to the emergence of speech by prediction of mother's responses to infant speech. Ultimately, our scientific paradigm likewise emerges by neurointeractivity as we learn to see the world in a way that explains more of the effects of our actions.

1.2   Introduction

A dominant trend in neuroscience research aims to characterize the sensory receptive field properties of cortical neurons. The success of this research is evident in the host of functional subdivisions across neocortex that have been defined this way. Much of this physiological research is based upon the method of transfer functions, a reverse engineering technique, which is used to characterize the output of an unknown device as a function of its inputs. This method correlates the neuronal activity observed in an area of cortex with respect to its sensory inputs or motor outputs. This correlation is used to define the function of each area with respect to the common field properties of its neurons as if that area is simply a transmission node in a communication pathway where its inputs are encoded and transformed into outputs.

Although highly productive, this analytical method implies an ill-founded "representation" paradigm, which views behavior as a reaction driven by inputs. This view has led to the predominance of the "neural code" concept and the idea that the function of a neuron (or cortical area) with a given receptive field (or class of fields) is to "represent" that characteristic of the input stimulus to the rest of the nervous system. The value of such a representation paradigm for an understanding of higher function is severely limited because it does not account for how the system must adapt to the representation and act upon unpredictable circumstances under the survival demands of dynamic environments. The representation paradigm is inherently open-ended as it defers the explanation of higher function to some sort of higher representation of representations without considering the path from output back to input that engages the environment. And it fails to explain how creative or autonomous behavior is generated without implying that a hidden supervisor or homunculus is responsible for interpreting and acting upon the representations.

In contrast, a comprehensive explanation of higher cortical function can be based upon a view of the cortex as part of a closed system, which engages environmental dynamics with the same interactive principles that govern its internal dynamics. Unlike the engineering approach, which must isolate a circuit element to determine its transfer function, a comprehensive approach must deal with the dynamical inter-activity generated collectively by reciprocally connected and mutually dependent cir-cuits. Cortical outputs influence all sources of cortical inputs (e.g. motor movements cause sensory motion) and alternative methods are necessary to deal with the extreme complexity that arises from such interactivity. This pervasive feedback encloses the ecological system of the behaving organism within the functional architecture of interactivity between cortical areas and the environment.

The "neurointeractive paradigm" views the behaving system with respect to the interwoven levels of organization and multiple time scales embedded within the cor-tical architecture (Figure 1.1). Local neuron interactivity is embedded within the interactivity of more distant connections between a multitude of cortical areas. This intra-cortical interactivity is embedded within the interactivity between the motor/sensory structures at the bottom of the cortical architecture, and between the neocor-tex and autoassociative structures at the top (i.e. hippocampus and prefrontal areas). And the closed system architecture seamlessly fuses this cortical network with the motor/sensory interactivity between the organism and the environment (or between communicating organisms). From this neurointeractive perspective, higher function emerges from the system as a whole by the dynamical mechanics of self-organization over a lifetime of continuous development within a mutual ecology that integrates the complexity of the organism with the complexity of the environment.

The neurointeractive paradigm avoids the homuncular pitfalls of sensory coding, representation and attention by emphasizing proactive sensory behavior rather than passive sensory information processing: listening rather than hearing; touching rather than feeling; looking rather than seeing. By placing the emphasis upon action, all conscious sensory behaviors are based upon the act of attending. This perspective implicitly avoids the difficulty of defining an attention mechanism without implying a supervisor that somehow knows what is important and what may be dissected from the total sensory representation. Instead, each and every action attends to a subjective prediction by testing the current associative hypothesis about the sensory conse-quences of that action. Such sensory behavior is proactive because the motor action and its associated pattern of predictive cortical activity precedes the sensory conse-quences of that action and is modified post hoc by those consequences. From this perspective, the primary function of the sensory systems is to provide feedback in a form that guides the next action toward the next test of one's subjective model of the world. This means that what one sees is largely determined by what one is looking for, rather than some sort of transformation of the objective world. The biological imperative of these interactive behaviors is to minimize uncertainty within a dynamic environment by learning to predict the sensory consequences of one's actions and by continuously testing those predictions.

This chapter will identify the fundamental principles of cortical organization and motor/sensory interactivity, which subserve the development of these action/predic-tion associations. Nonlinear dynamical systems analysis is the most appropriate description of cortical complexity and the vocabulary of this analysis provides a heu-ristic explanation for the emergence of higher function when applied to cortical neu-rointeractivity. The neurointeractive cycle of give and take between the cortical motor/sensory areas and the environment will be related to the dynamical mechanics of proactive exploratory behavior. By examining the process of early development with respect to the dynamical mechanics of cortical self-organization, the neurointeractive paradigm provides explanations for the emergence of higher functions such as self-identity, object recognition and speech communication.

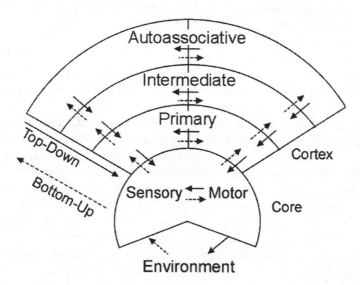

Figure 1.1. Closed hierarchical architecture of cortical interactivity. This highly simplified diagram represents the general organization of reciprocal bottom-up and top-down interconnections between the three essential shells of the cortical hierarchy. Areas within lower shells (i.e. Primary and Intermediate) provide bottom-up projections to the areas within higher shells (i.e. Intermediate and Autoassociative, respectively). Simultaneously, the higher areas provide top-down projections to areas in shells just below them. While a minimum of three such hierarchical shells are characteristic of neocortex, there are many areas within each of these shells and sub-shells may be distinguished, especially in primates. The entire cortical system may be subdivided into sensory and motor halves based upon the direct connections of the areas in the Primary shell to sub-cortical structures in the Sensory-Motor core. But interconnections between areas within each shell diffuse the sensory or motor identity of cortical areas throughout higher shells of the hierarchy. The self-sustaining cortical interactivity generated by this system of reciprocal interconnections reaches out to the environment through the motor and sensory structures in the sub-cortical core of the central nervous system for interactivity with selected elements of the Environment. Such extreme simplification is required to appreciate the closed nature of the system, and thereby eliminate the implication that some unidentified influence from outside the system is responsible for its complex behavior.

1.3   Principles of Cortical Neurointeractivity

The conceptual framework for this neurointeractive explanation of emergent higher function should be based upon the functional architecture of cortex where lesions have the most direct effects upon higher function. Beyond the enormous associative capacity of such extensively connected neural networks, at least five relatively unappreciated characteristics of cortex are essential for the dynamical neurointeractivity that subserves the emergence of higher function:

1. All areas of the cortex are always active. Although cortical states may be associated with fundamentally different modes of activity (i.e. slow-wave sleep versus alert desynchronization), neurons throughout cortex are always active.

2. The great majority of inputs to cortical neurons originate within the cortex itself. Almost all of the remaining inputs are from the thalamus, but the cortex is the major source of inputs to the thalamus. Therefore, with respect to the cerebral cortical system, which should be considered a network of interacting cortico-thalamic circuits, almost all connections in cortex provide cortical feedback.[1]

3. Outputs from a wide expanse of cortex, including the motor areas and the primary and higher order sensory areas, project directly to the subcortical structures responsible for the directed head, eye and finger movements that generate sensory inputs (e.g. cortico-tectal, cortico-pontine, cortico-spinal). Indeed, most sensory systems require movements or some other form of stimulus dynamics for the generation of receptor or primary afferent activity. This top-down control over the generation of bottom-up inputs is directly responsible for the proactive motor/sensory behavior that leads to the neurointeractive emergence of higher function. In addition, sensory cortex directly influences the earliest sensory nuclei in the central pathways (e.g. cortical projections to the dorsal horns or dorsal column nuclei directly influence somatosensory inputs where they enter the central nervous system). By all accounts, all sensory inputs that reach the cortex are influenced by top-down projections from the cortex itself.

4. All cortical connections are reciprocal. These reciprocal connections are the anatomical basis for cortical neurointeractivity. For instance, the primary sensory areas send "bottom-up" projections to secondary sensory areas (e.g. from visual area 17 to areas 18 and 19). These bottom-up projections are reciprocated by "top-down" projections from the secondary areas to the primary areas (e.g. from areas 18 and 19 to area 17). This directionality of the sensory path defines the cortical hierarchy from the primary areas,

---

[1] These two characteristics lead to the conclusion that cortical activity is self-sustained by pervasive endogenous feedback, especially in utero or during sensory deprivation or sleep. The term cortical "neurointeractivity" refers to this self-sustaining ensemble of collective neural behavior. Furthermore, given the plasticity of cortical synapses, cortical circuits and the neurointeractivity they sustain are in a continuous state of modification and self-organization.

which are direct targets of sensory inputs, to the higher order areas which receive indirect sensory inputs through the primary areas. In both directions, these cortical projections are always excitatory and it is likely that all areas may be activated by either bottom-up or top-down inputs. However, this top-down influence over cortical activity has not been thoroughly studied because the conventional transfer function approach would require careful manipulation of the top-down inputs, which are nearly inaccessible. The potential significance of this top-down activation of primary sensory areas is indicated by the finding that the primary visual areas may be activated in humans during mental imagery when the eyes are closed (Kosslyn et al., 2001). It is not useful to refer to either bottom-up or top-down projections as "the feedback" pathway because they are both sources of feedback with respect to the other. The dynamical complexity of cortical neurointeractivity is generated by this pervasive reciprocity, which extends across all areas of the cortex, creating a richly embedded system of multiple time-scales and interwoven levels of organization (i.e. interactivity between reciprocally connected primary sensory and motor areas is embedded within the interactivity between primary and secondary shells). The complexity of this collective neurointeractivity rises above the sum of all the embedded activities.

5.   The reciprocal connections between cortical areas are structurally and functionally asymmetric with respect to the spatial distribution of the axon projections, and with respect to the synaptic physiology of the connections (Figure 1.2).

Structurally, bottom-up projections from lower areas preserve the sensory topography with dense (<0.5 mm^2), point-to-point terminal axon clusters in the higher areas. In contrast, top-down projections from higher areas are distributed widely across the lower areas with top-down axons extending horizontally (> 2 mm) in all directions across the sensory topography in the lower areas (Rockland and Virga, 1989; Felleman and Van Essen, 1991; Cauller, 1995; Cauller et al., 1998). In addition, while the bottom-up cortical projections target a specialized population of local circuit neurons in middle layers, the top-down projections excite the subset of cortical neurons with layer I dendrites, which includes the large pyramidal cells that project to the brainstem nuclei for the top-down control of sensory-oriented movements.

Functionally, bottom-up projections activate a strong, but brief excitatory response (< 10 ms) which is abruptly terminated by a strong, long-lasting inhibition which may last longer than 50 ms (Douglas and Martin, 1991; Borg-Graham et al., 1998; Amitai, 2001). In contrast, the top-down projections activate a strong, relatively long-lasting excitation (> 30 ms) without inhibition (Cauller and Connors,

1994). Figure 1.3 relates the cortical microcircuit (Douglas and Martin, 1991) to the functional asymmetry of reciprocal cortical connections in the context of cortical neurointeractivity.

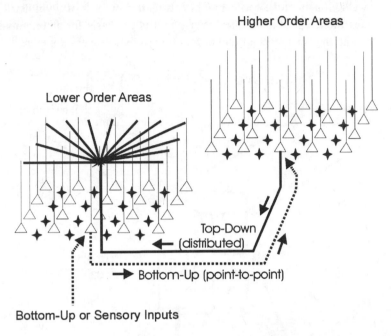

Figure 1.2. Asymmetric reciprocal connections between cortical shells. Areas within subjacent shells of the cortical hierarchy are represented as single layers of inhibitory (solid stellates) and excitatory (open pyramids) neurons. Figure 1.3 presents the more complex, multi-layered view of cortex. Bottom-up projections from lower shells of the hierarchy (i.e. sensory inputs or projections from primary areas to higher areas) excite both inhibitory and excitatory neurons within a relatively small locus of the higher cortical area. In contrast, the top-down projections from areas in higher shells ascend through the cortical layers to contact specialized dendrites of the excitatory neurons that extend to the surface, out of the reach of the inhibitory neurons. While each point in a lower area sends bottom-up projections to corresponding points in higher areas, the top-down projections from a point in a higher area extend horizontally across the surface of the lower area, which effectively distributes the higher influence to all points of the lower topography.

This reciprocal asymmetry has important functional consequences for cortical interactivity. This asymmetry superimposes topographic precision with widespread associativity throughout the cortical system. The distributed and long-lasting top-down projections generate an associative context throughout the cortical architecture which is sustained by the neurointeractivity generated by the higher order autoassociative areas. In contrast, the bottom-up pathway rapidly projects intense, spatially and temporally precise sensory patterns, which are immediately followed by equally precise and much longer lasting inhibitory after-patterns. This secondary bottom-up inhibition is the important source of sensory negative feedback, which resets the cycle of motor/sensory interactivity and propels the system toward an orthogonal attractor state. This process of motor/sensory interactivity interferes (positively and negatively) with the associative context that cascades over the cortical hierarchy from the top-down. Throughout the cortical architecture, this process of asymmetric interference drives the cycle of neurointeractivity that is the basis for autonomous self-organization and the emergence of higher function in dynamic environments.

Subjective Model of the World

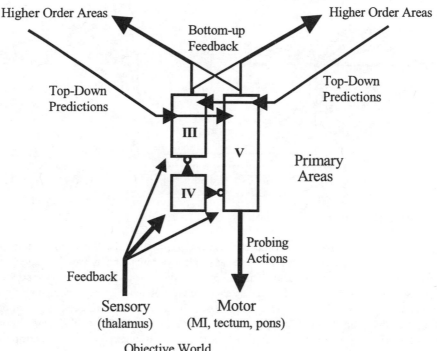

Figure 1.3. Canonical microcircuit for cortical neurointeractivity. The principal targets of the excitatory bottom-up projections from the sensory systems to the primary areas, and from lower to higher cortical areas, are the excitatory and inhibitory neurons found in the middle cortical layers. This initial bottom-up excitation spreads across cortical layers, but is abruptly terminated by the secondary inhibition generated by the response of the middle layer (IV) inhibitory neurons. Both the bottom-up and top-down projections between cortical areas originate primarily from the neurons in upper cortical layers (III), while the projections out of the cortex to motor and other sub-cortical structures originate from the neurons in lower cortical layers (V). In contrast to the middle layer neurons, which are most sensitive to bottom-up inputs, the projection neurons in upper or lower layers have specialized dendritic branches that extend to the surface of the cortex where the excitatory top-down projections from higher areas terminate. These top-down projections to the most superficial layers avoid the inhibitory neurons of the middle layers, such that top-down excitation is terminated primarily by the secondary inhibition generated by the middle layer response to bottom-up inputs.

1.4   Dynamical Mechanics

Nonlinear dynamical[2] systems analysis provides a comprehensive description of the neural mechanisms that mediate neurointeractivity and leads to a heuristic explanation for the emergence of higher function by cortical self-organization. Dynamical systems analysis identifies hidden deterministic structure in the midst of hyperdimensional complexity. We have found that even the simplest reciprocal networks of model neurons, based upon biologically realistic neural mechanisms, can generate dynamical interactivity with fractal chaotic structure or impose that structure upon a background of random inputs (Jackson et al., 1996). We have coined the term "chaoscillator" to characterize such a minimal network of two reciprocally connected excitatory/inhibitory neuron pairs because, following small changes in connection strengths or input intensity, their self-sustained behavior suddenly switches from periodic to chaotic. It should not be surprising that the behavior of cortex, which consists of enormous numbers of interacting chaoscillators embedded within a closed architecture of interwoven levels of organization and multiple time scales, also generates the unpredictability and rich spectral content of chaotic complexity.

From the perspective of the neurointeractive paradigm, the dynamical system of self-organized attractors that results from adaptive cortical reciprocity is the new functional equivalent to Hebbian cell assemblies. Indeed, the notion of the "phase sequence" (Hebb, 1949) or the "synfire chain" (Abeles, 1991) is literally equivalent to a dynamical attractor. Such attractors are patterns of activity that follow a deterministic path through the space of all possible activity patterns. The advantage of the attractor description is that it relates on-going spatiotemporal activity to the hyperdimensional structure of the system's history in the same way that planetary motion is related to the solar system of orbits within the galaxy of solar systems. This dynamical approach provides a higher dimensional perspective that helps to explain system behavior by identifying the geometry and the limits of its complexity.

There are several neural mechanisms that determine the path and cohesion of cortical attractors:

1.   Neurons are nonlinearly sensitive to coincident inputs (i.e. Abeles' synchronous gain). This nonlinearity turns each neuron into a coincidence or synchronicity detector such that the collective pattern of cortical activity is directly related to the pattern of correlation within the antecedent state of cortex.

2.   Synaptic feedback mediates both positive correlation by mutual excitation and negative correlation by mutual inhibition, both of which participate in the self-organization of the attractor structure by the nonlinear sensitivity of neurons to correlated inputs.

---

[2] The term "dynamical systems" is used to distinguish the specific case of dynamics that leads to temporal chaos and fractal geometry much like the term "classical" refers to the specific case in classical mechanics or classical music.

3. Winner-take-all domains, such as cortical columns, are local
    attractors, which may be generated by local cooperativity (i.e.
    mutual excitation or common inputs) embedded within a field of
    lateral inhibitory competition. The winner-take-all attractor phe-
    nomenon may also result from simple physical constraints. For
    instance, it is only possible to move the eyes or the arm in one
    direction at a time. Any such winner-take-all phenomenon con-
    strains a specific subset of correlated activities.

4. Long-term synaptic plasticity strengthens attractors by associativ-
    ity, both by enhancing the connections between coactive neurons
    (i.e. Hebb's postulate), and by depressing the connections
    between neurons whose activity is uncorrelated (i.e. anti-Heb-
    bian). Such associative neurophysiological mechanisms may
    result in nearly permanent changes in connection strength and are
    widely believed to be responsible for memory. These associative
    mechanisms tend to stabilize the system-wide attractor structure
    because they lead to repetition, and repetition further strengthens
    the attractor structure. This plasticity is synergetic with the other
    attractor mechanisms, which also depend upon neural correlation.

All such attractor mechanisms carve out and strengthen the specific path that the
system behavior travels through the state space of all possible behaviors. In general,
these neural attractor mechanisms are excitatory. But the ubiquitous inhibition that
keeps check against runaway excitation plays an equally important role for attractor
formation by inhibiting states (e.g. "losers") that are incompatible with the attractor.
These attractor mechanisms are the gravity and glue that hold together the dynamical
structure of neurointeractivity by pulling the system behavior toward quasi-stable
fixed-points or periodic limit cycles.

The unpredictable richness of deterministic chaos can be described in terms of a
"strange attractor" that temporally escapes a quasi-periodic orbit and re-enters with
different initial conditions, travels through a higher or lower orbit, and then escapes
again, re-enters, and so on. The escape is mediated by "repeller" mechanisms, which
create a saddle point in the orbit that destabilizes the system away from the attractor.
By contrapuntal interplay, such coupled attractor/repeller mechanisms propel the system to autonomously explore its state space and self-organize a system-wide, associative hyper-structure of quasi-stable attractor sequences (Figure 1.4).

Several neural mechanisms mediate the repeller dynamics of neurointeractivity:

1. Intrinsic neuronal outward currents such as the voltage-sensitive
    IA or calcium-dependent IK(Ca) mediate frequency adaptation,
    after-hyperpolarization or other types of intrinsic inhibition that
    are secondary to strong, sustained activation.

2. Short-term synaptic depression decreases the strength of connections from neurons that remain highly active (Markram, Chapter 5, this volume).

3. Local inhibitory feedback is greatest for those neurons that are most active because they are at the central focus of all the surrounding inhibitory neurons excited by that activity. This mechanism is a ubiquitous characteristic of neural systems. While its role in controlling runaway excitation is usually emphasized, this repeller function of recurrent inhibition employs the same mechanism to propel the system behavior from one action to the next.

4. The long-range feedback inhibition mediated by the secondary component of the cortical response to bottom-up sensory inputs imposes a negative after-image with the same, high spatial resolution of the bottom-up sensory topography. This repeller corresponds to the negative sensory feedback that terminates the current action and guides the attractor sequences for exploratory behavior.

All such repeller mechanisms are secondary to the activation generated by an attractor. They grow in strength during that activation until they shut down that particular attractor and allow the system to escape to another attractor. The persisting repeller effects generated in response to an attractor create an anti-attractor (like a negative after-image) that drives the system away from that particular attractor toward the specific subspace of pseudo-orthogonal attractors that are normally inhibited by the attractor.

This coupling between attractor and repeller mechanisms constrains the sequential order of quasi-stable attractor states and segments a hyper-space of potential attractor sequences. The complexity of a particular attractor sequence (i.e. series of eye movements) is a neurointeractive product of the environmental consequences of that sequence on the one hand (i.e. series of visual signals generated by looking at an object), and of the hyper-structure of the system-wide associative context on the other (i.e. what one thinks they are looking at). Depending upon the bottom-up sensory consequences of each action, the top-down influences of the associative context bias the primary motor/sensory areas toward a specific next attractor action within the space of all orthogonal attractors. In this way, the positive (or negative) interference between the associative context and the motor/sensory interactivity with the environment simultaneously guides the attractor sequence and strengthens (or weakens) the associative hyper-structure by correlation (or decorrelation).

The associative context ties together the action/prediction associations in the form of motor/sensory attractors related to a specific set of environmental contingencies such as those encountered while interacting with an object or while communicating with a person. Over the unique course of development, a subjective model of the world grows as a strange attractor hyper-structure of associative contexts. An associative context, which is self-sustained by the auto-associative neurointeractivity at the

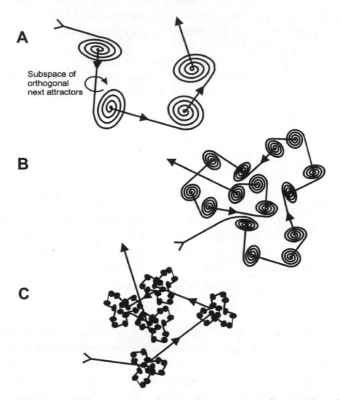

Figure 1.4. Self-similar hyper-structure of cortical attractor/repeller saddle points. (A) The path of cortical activity through state space falls toward attractors represented here as spirals approaching fixed points to provide a specific locus for the saddle point where the repeller overcomes the attractor. But such saddle points may occur as the state approaches limit cycles or more complex attractors as well, which may themselves appear as fixed points from another hyper-dimensional perspective. As repeller mechanisms take over, the activity state is propelled from the saddle point toward an orthogonal set of attractors represented here as a trajectory that is perpendicular to the plane of the spiral. Such saddle points liberate the activity state to move from one elementary action attractor (e.g. sensory/motor movement) to an orthogonal set of next actions thereby maintaining a structure that links together action sequences. (B) At a larger scale in hyper-dimensional state space, represented here in only three dimensions, sequences of elementary actions (as in A) are linked together into a similar hyper-attractor represented here as a spiral of spiral sequences. An example of such a hyper-attractor is an associative structure (e.g. for interaction with a complex object such as a tennis ball) that extends beyond the primary areas to tie together sensory/motor movement sequences that are reinforced by successful sensory predictions. The saddle points of such associative hyper-attractors propel the activity state from one associative context to the next depending upon the higher order consequences of the actions (e.g. this is probably a squash ball because it doesn't bounce well). (C) This hyper-structural process is extended here in the form of a spiral of spirals of spiral sequences to illustrate that saddle points may exist across the scales of associative structures from elementary movements to complex conceptual frameworks. This hyper-structure can tie together all activity states throughout the cortex in a way that is guided by interactive experience.

top of the cortical hierarchy, imposes relative stability for the immediate motor/sensory actions in a dynamic environment. This higher order neurointeractivity changes relatively slowly because it is indirectly coupled to the rapid neurointeractivity in the motor/sensory areas during behavior. The long-lasting excitation and widespread distribution of these top-down associative influences impose a contextual background of structured activity. This context serves as a reference with which the immediate motor/sensory attractors become associated by the covariance rules of synaptic plasticity mechanisms. As a result, the associative context gains the power to bias the attractor sequence and promote a context-based series of actions, depending upon the sensory experiences that occur when that context is present.

These dynamical mechanics of cortical neurointeractivity generate a rich diversity of adaptive behavioral patterns and provide the necessary flexibility to cope with the survival demands of dynamic environments. These mechanisms propel behavior through the system-wide, associative hyper-structure of attractor sequences guided by experience and the environmental consequences of one's actions. Self-organization during environmental interactions reinforces action sequences whose sensory consequences stabilize the cortical neurointeractivity between motor and sensory areas within the hyper-structure of associative contexts that extends throughout. The embedded cortical system of interwoven levels of organization and multiple timescales generates rapid motor/sensory interactivity with dynamic environmental contingencies under the relative stability imposed from the top-down by the higher order associative context. This dynamical systems description of cortical neurointeractivity provides a comprehensive account of extreme complexity from a geometric perspective that helps to explain the hyper-dimensional emergence of higher function.

1.5   The Neurointeractive Cycle

The "neurointeractive cycle" (Figure 1.5) describes the dynamical cortical process that subserves exploratory behavior in terms of the attractor/repeller mechanics that generate adaptive sequences of quasi-stable attractors. While analogous cycles may describe the dynamical interactivity throughout the network of reciprocally connected cortical areas, the following description refers to the actions mediated by the motor and sensory areas of cerebral neocortex because these can be related most directly to observable behavior. The associative context guides this neurointeractive cycle, which generates specific actions and, in turn, the sensory consequences of those actions modify the associative context. The neurointeractive cycle drives the continuous process of dynamical self-organization and the immediate give and take of proactive behavior as it revolves around four functional phases:

1. *Top-Down Prediction.* The widely distributed top-down influence of the associative context biases the neurointeractivity between the motor and sensory areas toward a specific winner-take-all attractor. Prior to the cortical generation of any movement or other action, this cortical motor/sensory attractor associates activ-

ity in the sensory areas with the activity in the motor areas that generates the action. This cortical sensory component becomes predictive because it is modified by the sensory consequences of the motor action. The next time this motor/sensory attractor is activated, the sensory component will predict the consequences of the motor action. In this way, the quasi-stable dynamics of the attractor create a temporally inverted association between an intended action and its consequences.

2. *Probing Action.* The pattern of corticofugal activity within the motor component of the motor/sensory attractor generates a specific winner-take-all action or movement. This action pattern is a small facet of a hyper-dimensional complex associative structure. Any given action may be generated under a wide range of associative contexts, but each action probes the environment and tests a specific, top-down sensory prediction that is consistent with the current associative context.

3. *Bottom-Up Feedback.* The environmental consequences of the motor action generate a pattern of bottom-up sensory inputs. This bottom-up pattern rapidly ascends through the cortical hierarchy with characteristically high precision with respect to both the spatial topography of the sensory input and the timing of the excitatory sensory signal. Collision of this bottom-up signal with the top-down context triggers associative mechanisms throughout the cortex and deflects or reinforces the attractor path that generated the probing action. Secondarily, the inhibitory phase of this bottom-up sensory projection generates a negative after-image that repels the system away from the current motor/sensory attractor toward a new action that is associated with different sensory predictions of the same associative context. To the extent that the bottom-up sensory input pattern corresponds with the cortical sensory component of this motor/sensory attractor, the secondary bottom-up repeller mechanism enhances the intrinsic repeller of recurrent inhibition that follows all attractors and propels behavior to the next action/prediction.

4. *Associative Interference.* There are several effects of the bottom-up sensory feedback that guide the short-term evolution of the attractor sequence and the long-term self-organization of the associative hyper-structure. As the bottom-up sensory pattern ascends through the cortical hierarchy, it interferes with the endogenous pattern of activity sustained by the higher order associative context. To the extent that the cortical sensory pattern associated with the motor action predicts the pattern of sensory inputs generated by that action, associative correlation or "positive interference"

occurs. Nonlinear excitation (i.e. synchronous versus asynchronous coactivation) is generated at points of positive interference, which reinforces the motor/sensory attractor as well as the secondary repeller generated by that attractor. Similarly, at points where the sensory input pattern is correlated or uncorrelated with the cortical action/prediction attractor, the higher order predictive success of the top-down associative context is strengthened or weakened, respectively, by the associative mechanisms of synaptic plasticity.

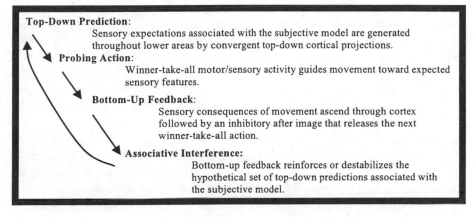

Top-Down Prediction:
   Sensory expectations associated with the subjective model are generated throughout lower areas by convergent top-down cortical projections.
Probing Action:
   Winner-take-all motor/sensory activity guides movement toward expected sensory features.
Bottom-Up Feedback:
   Sensory consequences of movement ascend through cortex followed by an inhibitory after image that releases the next winner-take-all action.
Associative Interference:
   Bottom-up feedback reinforces or destabilizes the hypothetical set of top-down predictions associated with the subjective model.

Figure 1.5. Interactive cortical cycle of active sensory hypothesis testing. The dynamical interactivity generated by these reciprocal projections between cortical areas generates a cycle of self-organization that is propelled by the mechanisms of the cortical circuit. These neural mechanisms support a regenerative, but non-repetitive cycle of cortical activity: the top-down influence of distributed cortical activity generates activity in lower areas, some of which produce motor actions; the consequences of this top-down activation and the motor actions it generates results in negative sensory feedback and other bottom-up inhibition; in turn, this bottom-up inhibition at specific cortical loci releases the next top-down motor action generated at other cortical loci under the continuing top-down influence of the more slowly evolving pattern of distributed cortical activity. This figure relates equivalent functions to each phase of this interactive neural cycle which emerge in the context of sensory/motor interactivity with the objective world to give meaning to both the structure of the distributed cortical activity and the overt actions associated with that structure. To the extent that this associative structure successfully predicts the bottom-up or sensory consequences of these actions, the structure is reinforced by reactivation of predicted cortical loci, or it is destabilized by bottom-up activation of unpredicted loci which moves the system toward a new associative structure.

This cycle is the driving force at the core of the neurointeractive process distributed throughout the cortical system. The cortex employs the same neurointeractive cycle that generates a variable repertoire of flexible behavior to explore environmental complexity, and simultaneously construct and refine the cortical associative hyper-structure that guides this exploratory behavior. As the cortex adapts and the sensory prediction improves, positive interference also strengthens the secondary repeller effect of the bottom-up sensory negative feedback by improving the coupling between the sensory input and the sensory prediction component of the action/prediction attractor. This drives the system away from the action/prediction attractor more effectively, prevents repetition, and propels the system toward a more distinct orthogonal subspace of attractors that emit a new probing action to explore a distinct set of predictions.

The top-down associative context simultaneously guides the sequence of action/prediction attractors, is modified by the bottom-up sensory consequences, and serves as a relatively stable associative reference for the action/prediction sequence as a whole (Figure 1.5). The neurointeractive cycle supports repetitive actions until the sensory component of the cortical action/prediction attractor adapts to predict the sensory consequences of these actions. As the prediction improves, the secondary bottom-up inhibition of the sensory consequences becomes aligned with the cortical sensory activity and strengthens the repeller to propel the system toward the next action, and then on to the next associative context. A sign of this adaptive process should be a transition from repetitive, highly stereotyped actions to more variable and experimental behaviors. Conversely, if the associative context does not account for the sensory consequences of its exploratory actions, the repeller remains too weak to propel the system through its dynamical structure. In this case, the neurointeractive cycle generates repetitive, persistent exploratory actions until the system learns to predict the sensory consequences or the associative context destabilizes in the absence of correlative support and another context takes over and is tested for its ability to account for its actions. In this way, the exploratory behavior of neurointeractivity steadily makes better predictions, probes more effectively, and naturally discovers the rules of environmental contingencies.

Another way of looking at this neurointeractive cycle is in terms of experimental hypothesis testing. The neurointeractive cycle solves the inverse problem of coping with extreme environmental complexity using severely limited sensory and motor abilities by a process that is analogous to the scientific method: Theory leads to specific hypothetical predictions, experimental actions, and comparison of predictions with measured observations, which, in turn, leads to confirmation, refinement or modification of theory, further experimentation, and so on. The top-down associative influence over the motor/sensory areas generates an elementary hypothesis about the nature of the environment, which predicts the specific sensory consequences that should follow a specific motor action. Then the interference between the top-down sensory prediction and the bottom-up sensory consequences of that action test the suitability of the hypothesis in a form that directly modifies the associative structure, or thesis, that generated the hypothesis. In an analogous way, our scientific view of the objective world is modified by experimentation. Indeed, our subjective experience of the world around us may be drastically changed by scientific discovery. Accordingly, scientific method may be considered a formalization of the neurointeractive process that generates conscious behavior.

Other theories have also emphasized the basic process of hypothesis testing. Most of these have described sensory processing in terms of a prediction generated from the top-down in the primary sensory cortex (e.g. Mumford, 1994; Ullman, 1994).

However, such predictions are typically in the form of a static, complete image that is compared to the bottom-up sensory pattern without considering the movements or other sensory-oriented actions that are essential for dynamical neurointeractivity and exploratory behavior. Although such models also view the cortical process in terms of a solution to the inverse problem of sensory perception, they remain basically representational because their goal is to reconstruct an accurate image of the sensory pattern, usually without specifying how such a representation is interpreted or related to behavior.

There is no need to reconstruct the structure of the environment in the form of a complete sensory image. Most objective details, such as the texture of wood or the color of water, are not essential to behavior and they may be re-observed directly, in all their complexity for fascination at some other time. Indeed, the fleeting sensory images generated during conscious behavior are sparse and incompletely processed (Quartz and Sejnowski, 1997, *ibid*). Instead, it is only necessary to predict the elementary sensory consequences of a specific subjective action that are critical tests of the current associative hypothesis (e.g. rightward eye movement across a hypothetical flat square should encounter two equal parallel edges, which fade before the next, probably perpendicular movement). The neurointeractive process emphasizes these simple predictions of sensory elements in direct relation to elementary actions, rather than the reconstruction of an accurate representation of the sensory image. Such a process of experimentation is highly robust with respect to generalization from specific experiences to other perspectives and variable conditions, because it actively searches for pieces of supporting evidence. When such hypothesis testing models are extended to describe the multitude of reciprocal interactions throughout the closed cortical system, an extremely powerful associative process is combined with the ability for autonomous exploratory behavior.

The "hierarchical clustering" model developed by Ambros-Ingerson et al., (1990) provides a more neurointeractive form of hypothesis testing which also provides a solution to the inverse problem of perception. Their model categorizes complex sensory patterns by generating a sequence of increasingly specific subcategories through a reiterative cycle that removes the features of the less specific categories from the input pattern. With respect to dynamical mechanics, neurointeractivity within this network generates a specific sequence of pseudo-orthogonal, local winner-take-all attractors, propelled by top-down inhibitory repellers that are coupled to the attractors. In many ways, the neural basis of this categorization process is analogous to the neurointeractive cycle.

Such sensory categorization models have primarily been applied to the analysis of a static sensory input pattern without consideration of the interactive behavior that is essential for most conscious sensation. The hierarchical clustering model emphasizes neurointeractivity between the cortex and the peripheral olfactory structures, and such models in general are well suited for olfaction where a static sensory pattern may be assumed.[3] Like the "sparse coding" model of primary sensory processing

---

[3] However, it is probably significant that in rodents where this hierarchical model has been applied most specifically, sniffing movements are synchronized with whisking movements that probe the somatosensory domain at the same time.

developed by Olshausen and Field (1997), such categorization models demonstrate how associative correlation may generate a statistical components analysis of complex input patterns. The sparse coding model emphasizes neurointeractivity between primary and secondary cortical visual areas, but like other categorization models, it does not deal with behavioral interactivity or environmental dynamics. In many ways, the neurointeractive process of exploration and discovery is analogous to an extended version of these categorization models. For instance, it may be possible to demonstrate that the neurointeractive cycle generates a dynamical sequence of action/prediction attractors during exploration in natural environments that progresses hierarchically from general impressions to critical tests of specific hypotheses generated by the associative context. Such a comprehensive model of hierarchical neurointeractivity would provide robust flexibility because it shifts the center of immediate action outward to directly engage the dynamics of environmental complexity.

From the neurointeractive perspective, action/prediction attractor sequences related to specific experiences are tied together by an associative context, and the dynamical interdependence of these contexts as a whole forms a system-wide associative hyper-structure which adapts to environmental complexity. This hyper-structure becomes a uniquely subjective theoretical framework or paradigm or model of the world, which ties together all the associations between specific motor actions and sensory expectations. These associations define the meaning of behavioral actions in terms of the subjective experiences that stabilize the action/prediction attractors. Similarly, the neurointeractive system's model of the world is purely subjective. It is built upon the associative structure that was self-organized from the beginning of development, for a large part in the absence of structured sensory inputs, from the unique perspective and history of the organism. This subjective model of the world is the neurointeractive alternative to the sensory transformation of the objective world that characterizes the representational paradigm. By dynamical self-organization, the neurointeractive cycle incrementally constructs this subjective model out of the same unique patterns of cortical activity that mediate the continuous flow of conscious exploratory behavior.

1.6   Developmental Emergence

Given the self-sustaining predominance of cortical feedback and ubiquitous synaptic plasticity, the associative structure of cortical activity continues to self-organize from the first moments of cortical development in utero. Throughout the closed cortical system, the dynamical mechanics of cortical neurointeractivity pull this associative structure up by the bootstraps. The prenatal development of lateral inhibition and other attractor mechanisms provides the anatomical basis for local winner-take-all domains and the greater complexity of interwoven functional columns. Even in the absence of sensory inputs, the development of cortical attractors associates patterns of activity in the motor areas with patterns in the sensory areas, which directly interconnect with those motor areas. The newborn infant spends most of its time in rapid eye movement (REM) sleep and many have wondered about the sensory content of a newborn's dream. From the neurointeractive perspective, this REM sleep corresponds to the process of dynamical self-organization that prepares the infant to explore environmental complexity.

From the first moments following birth, environmental sensory consequences become contingent upon motor actions (e.g. hearing oneself cry, seeing the visual motion generated by head and eye movements, or feeling the texture of cloth when the limbs are moved). This newborn behavioral/environmental interactivity joins the cortical motor/sensory interactivity embedded within the self-organized complexity of the newborn's associative structure.

As associative neural mechanisms depend upon input correlation, attractors that associate the most reliable sensory consequences with movements or other actions will develop first. By far, the most reliable sensory consequences are those that are directly generated by one's own actions in a static environment (e.g. rightward visual motion with leftward eye movements, or varying sounds with modulated vocal action). As the newborn continues to interact with the environment, these fundamental action/prediction attractors will strengthen and form a durable, subjective framework for the associative hyper-structure that grows with the system throughout its lifetime. All higher action/prediction attractors are built upon this framework of "self" complexity. And the neurointeractive process that constructs the subjective model of "self identity" leads to the emergence of higher function as the self becomes distinguished from other, and then grows to include one's influence upon objects and those significant others who respond reliably to our actions.

This emergence of self identity may be specifically related to the dynamical mechanics of the neurointeractive cycle. As the cortex gains influence over the actions of the newborn, the pattern of motor activity that effects movements is already associated with internally consistent patterns of cortical activity in the sensory areas, which have developed without the correlative influence of the sensory input systems. But once the cortical motor activity generates reliable sensory consequences, the associative interference phase of the neurointeractive cycle aligns the sensory component of the cortical motor/sensory attractor with these highly predictable input patterns. Soon, given the reliability of these predictions in nurturing environments, the sensory component of the motor/sensory attractor predicts the sensory consequences of the motor action. The next time a pattern of cortical motor activity generates a probing action, it is associated with a pattern of cortical sensory activity that predicts the sensory consequences of that action, even before the action is emitted. In this way, the cortical attractor that generates rightward eye-movements simultaneously predicts leftward visual motion and vertical edges as part of its self-consistent associative structure. Similarly, the auditory cortical pattern for high pitched sounds becomes an integral component of the motor/sensory attractor that tightens vocal tension. These motor/sensory attractors are the first to stabilize in the newborn under the reliable feedback from the sensory system, and they are reinforced throughout a lifetime of interacting with the environment.

The same neurointeractive process gradually aligns the higher order associative hyper-structure with these stable motor/sensory attractors and creates a subjective model of self identity, which ties together the set of all attractors that predict the self-consistent sensory consequences of probing actions (e.g. rightward sensory consequences of leftward action). By providing a widely distributed, top-down referential background for feature analysis, self identity provides a stable framework for object identification and the construction of a subjective model of the objective world. As the infant probes objects in the environment, the physical characteristics of those objects generate consistent sensory features that are highly correlated with the probing actions. For instance, horizontal eye movements across a square generate a parallel pair of equal length edges out of all the possible vertical edges that are self-consistent with that action. The square object is further identified by subsequent vertical eye movements that generate a correlated subset of horizontal edges. In the presence of a specific object, the neurointeractive cycle modifies the top-down predictions of those probing actions by the associative interference between elements of the self-consistent framework and the highly correlated set of sensory consequences generated by the object. In this way, the associative context related to a specific object is created out of the structure of self-identity as it guides a sequence of probing actions, which, in turn, become associated with a subset of self-consistent predictions. By generating probing actions that critically test hypotheses about the unique identity of the object, the associative context distinguishes the object from the background of self.

Speech and other vocal communication are very special cases of neurointeractivity. The development of speech in humans is perhaps the most carefully studied form of cognitive development and has already been described in terms of dynamical systems analysis (Elman et al., 1997). Although characteristically variable in the specific timing of developmental stages, the sequential growth of infant speech complexity is well established. In particular, with respect to the neurointeractive cycle, infant babbling progresses from reduplicated to variegated sequences of elementary speech action (i.e. bababa to baba-dada) before it progresses to referential word formation. Accordingly, the neurointeractive cycle would generate repetitive speech actions until the cortical sensory component of the motor/sensory attractor adapts to match the sensory consequences of those actions. As this prediction improves, the secondary inhibition generated by the bottom-up sensory inputs becomes aligned with the action/prediction attractor, thereby strengthening the repeller and propelling the system toward a new action/prediction attractor. This cortical neurointeractivity generates the variable sequences of speech sounds observed in variegated babbling. And the development of babbling behavior provides a revealing demonstration of the neurointeractive formation of self identity as the infant learns to predict the sounds of its own speech.

A neurointeractive explanation for the development of speech communication is especially interesting because the cortical activity in the sensory areas, which adapts to predict the sensory consequences of one's own speech actions, is simultaneously prepared to predict those similar sensory patterns generated by the speech of others. Such communication is extremely effective because it is directly constructed from the subjective framework of self-consistent action/prediction associations. The caregiver guides the development of this communication by nurturing a bridge from the babbling speech of the infant, to the complexity of language. This "motherese" approximates the infant's speech with an exaggerated prosody that segments and emphasizes the language elements such as subject, action, object (e.g. mama loves baby). Motherese interacts with the same action/prediction attractors that are self-organized during babbling and creates an associative context that biases the speech attractor sequence toward the language structure. This establishes an almost direct, intimate link between the cortical neurointeractivities of the speaker and the listener.

The neurointeractive cycle continues to build upon the associative hyper-structure by learning to predict the consequences of one's actions upon objects and others by the same process that self-organizes the framework of self identity. Manipulative actions upon objects generate sensory consequences that create cortical predictions. The structure of these action/prediction associations adapts to the complexity of the environmental physics. Similarly, the associative hyper-structure adapts to predict the influence of one's speech upon the behavior of others, and the structure of these action/prediction associations adapts to the complexity of the language. In this way, the framework of self grows by incorporating one's predictable influence upon the environment.

1.7 Explaining Emergence

A direct, explicit explanation, like the explanation of planetary movements or the explanation of chemical reactions, is not possible for the explanation of emergent properties such as the hydrodynamics of H_2O or the economics of the market place or the autonomy of conscious behavior. In contrast to systems that are driven by global external forces such as gravity or thermodynamics, emergence arises from complex systems whose dynamics are dominated by the local interactivity between extremely large numbers of elements. Given this pervasive interdependence, an explicit account of the behavior of any single element would require a recursive accounting of all the elements. This complexity and enormity is beyond the capacity for a comprehensive understanding even though the individual elements and interactive physics may be relatively simple. Although it may be impossible to map direct causal relations for a specific emergent phenomenon, scientific methods may determine the relations between emergent properties that apply to the higher order physics (i.e. the thermodynamics of chemical reactions or the psychology of neural systems). At some point, the explanation of emergence depends upon a quantal leap in understanding from the structure and dynamics of the elements, to the functions and interrelations of the collective properties.

Computer modeling techniques that make it possible for the first time in history to realistically simulate the collective behavior of enormous numbers of simple elements (i.e. weather patterns or brain activity) offer the best tool for the analysis of complex systems. The growing field of dynamical nonlinear systems analysis provides the most appropriate description of this complex behavior. However, methods for demonstrating the emergence of higher function using such computational or mathematical models face obstacles that are not generally appreciated. Functionality is not pre-programmed, so a long period of autonomous interactivity in a complex environment is required for the dynamical self-organization of adaptive behavioral patterns (e.g. infants take years to develop speech). Particular functions (e.g. speech or visual recognition) emerge from dynamical system properties (e.g. communication and exploration) by real-time interactivity with the specific environmental contingencies that nurture those functions. Therefore, to demonstrate emergence using computational models, the nurturing environment, either real or virtual, must be included in the model for the higher function to be recognized. Unfortunately, a comprehensive simulation of the nurturing environment may be more difficult to construct than an artificial nervous system. And an artificial nervous system capable of real-time interactions with real environments requires extremely advanced computer and robotic hardware. These technical breakthroughs are imminent, but a major shift toward the neurointeractive paradigm is necessary before such tools may help us explain the emergence of higher function.

References

Abeles, M. (1991) Corticonics: Neural Circuits of the Cerebral Cortex. Cambridge, UK: Cambridge University Press.

Ambros-Ingerson, J., Granger, R., Lynch, G. (1990) Simulation of paleocortex performs hierarchical clustering. Science 247: 1344–1348.

Amitai, Y. (2001) Thalamocortical synaptic connections: Efficacy, modulation, inhibition and plasticity. Rev Neurosci. 12 (2): 159–173.

Borg-Graham, L.J., Monier, C., Fregnac, Y. (1998) Visual input evokes transient and strong shunting inhibition in visual cortical neurons. Nature 393: 369–373.

Cauller, L.J. (1995) Layer I of primary sensory neocortex: Where top-down converges upon bottom-up. Behavioural Brain Research 71: 163–170.

Cauller, L.J., Clancy, B., Connors, B.W. (1998) Backward cortical projections to primary somatosensory cortex in rats extend long horizontal axons in layer I. J. Comp. Neurol. 390: 297–310.

Cauller, L.J., Connors, B.W. (1994) Synaptic physiology of horizontal afferents to layer I in slices of rat SI neocortex. J. Neuroscience 14: 751–762.

Cauller, L.J., Kulics, A.T. (1991) The neural basis of the behaviorally relevant N1 component of the somatosensory evoked potential in SI cortex of awake monkeys: Evidence that backward cortical projections signal conscious touch sensation. Exp. Brain Res. 84: 607–619.

Douglas, R.J., Martin, K.A. (1991) A functional microcircuit for cat visual cortex. J. Physiol. 440: 735–769.

Elman, J.L., Parisi, D., Bates, E.A., Johnson, M.H., Karmiloff-Smith, A. (1997) Rethinking Innateness: A connectionist perspective on development. Boston: MIT Press.

Felleman, D.J., Van Essen, D.C. (1991) Distributed hierarchical processing in the primate cerebral cortex. Cereb. Cortex 1 (1): 1–47.

Hebb, D.O. (1949) The Organization of Behavior. New York: John Wiley.

Jackson, M.E., Patterson, J., Cauller, L.J. (1996) Dynamical analysis of spike trains in a simulation of reciprocally connected "chaoscillators": Dependence of spike pattern fractal dimension on strength of feedback connections. In: J.M. Bower (Ed.) Computational Neuroscience: Trends in Research. San Diego: Academic Press.

Kosslyn, S.M., Ganis, G., Thompson, W.L. (2001) Neural foundations of imagery. Nat. Rev. Neurosci. 2 (9): 635–642.

Mumford, D. (1994) Neuronal architectures for pattern-theoretic problems. In: C. Koch, J.L. Davis (Eds.) Large-Scale Neuronal Theories of the Brain. Cambridge, MA: MIT Press, pp. 125–152.

Olshausen, B.A., Field, D.J. (1997) Sparse coding with an overcomplete basis set: A strategy employed by V1?. Vision Res. 37: 3311–3325.

Quartz, S.R., Sejnowski, T.J. (1997) The neural basis of cognitive development: A constructivist manifesto. Behav. Brain Sci. 20: 537–596.

Rockland, K.S., Virga, A. (1989) Terminal arbors of individual "feedback" axons projecting from area V2 to V1 in the macaque monkey: A study using immuno-histochemistry of anteogradely transported Phaseolus vulgaris-leucoagglutinin. J. Comp. Neurol. 285 (1): 54–72.

Ullman, S. (1994) Sequence seeking and counterstreams: A model for bidirectional information flow in the cortex. In: C. Koch, J.L. Davis (Eds.) Large-Scale Neuronal Theories of the Brain. Cambridge, MA: MIT Press, pp. 257–270.

Chapter 2

The Cortical Pyramidal Cell as a Set of Interacting Error Backpropagating Dendrites: Mechanism for Discovering Nature's Order

Oleg V. Favorov, Joseph T. Hester, Douglas G. Kelly, Dan Ryder, Mark Tommerdahl

2.1 Abstract

Central to our ability to have behaviors adapted to environmental circumstances is our skill at recognizing and making use of the orderly nature of the environment. This is a most remarkable skill, considering that behaviorally significant environmental regularities are not easy to discern: they are complex, operating among multi-level nonlinear combinations of environmental conditions, which are orders of complexity removed from raw sensory inputs. How the brain is able to recognize such high-order conditional regularities is, arguably, the most fundamental question facing neuroscience. We propose that the brain's basic mechanism for discovering such complex regularities is implemented at the level of individual pyramidal cells in the cerebral cortex. The proposal has three essential components:

1. Pyramidal cells have 5–8 principal dendrites. Each such dendrite is a functional analog of an error backpropagating network, capable of learning complex, *nonlinear* input-to-output transfer functions.

2. Each dendrite is trained, in learning its transfer function, by all the other principal dendrites of the same cell. These dendrites teach each other to respond to their separate inputs with *matching* outputs.

3. Exposed to different but related information about the sensory environment, principal dendrites of the same cell tune to different nonlinear *combinations* of environmental conditions that are *predictably related*. As a result, the cell as a whole tunes to a set of related combinations of environmental conditions that define an orderly feature of the environment.

Single pyramidal cells, of course, are not omnipotent as to the complexity of orderly relations they can discover in their sensory environments. However, when organized into feed-forward/feedback layers, they can build their discoveries on the discoveries of other cells, thus cooperatively unraveling nature's more and more complex regularities. If correct, this new understanding of the pyramidal cell's functional nature offers a fundamentally new insight into the brain's function and identifies what might be one of the key neural computational operations underlying the brain's tremendous cognitive and behavioral capabilities.

2.2   Introduction

When a student first becomes attracted to neuroscience, it is often because the great mystery of the mind-body problem beckons. "How could that wrinkly lump of cells produce...this!" he might exclaim. "It *must*, I know that, but how?" And he imagines the culmination of his career to be attaining just that understanding.

We all chuckle. Much later, we think, when he comes to appreciate the realities of single neuron physiology, anatomical tracing, genetic analysis, behavioral testing, and (perhaps most importantly) grant writing, he will realize the foolishness of his dream. Neuroscience is in the details. And there are too many details to hope for "understanding the mind/brain" (whatever that might be!) before many generations hence. And even then it will probably be a collective understanding, where Professor C. understands the vermis of the cerebellum, Dr M. gets a handle on the intricacies of primary motor cortex control of pianist finger movements, and the Y Institute has mapped out the important G-proteins to be found in a typical layer V pyramidal cell in V1.

Is this what we have to look forward to? Possibly, if the brain is, as is commonly believed, an amalgam of many different circuits constructed on different principles to carry out different functions. We should then expect that an understanding of how the brain works will be attained only after a painstaking process of identifying those circuits, discovering how they function, how they are built and, ultimately, learning how they work together to endow us with our remarkable mental and behavioral abilities.

Older scientific traditions in such disciplines as physics, chemistry, and biology stand in stark contrast to this vision of a balkanized neuroscience. These older traditions are unified and driven forward by *key theoretical insights*, insights that uncover fundamental properties which lie at the causal roots of a natural kind of phenomena. Therein lies their explanatory power. For example, the consistent causal structure that

DNA imposes allowed Mendel, Darwin, and Crick and Watson to furnish us with key biological insights. Similarly, the systematic structure of the atom allowed for the theoretical advances of Mendeleev and Bohr. Mechanics was a subject just asking to be described by basic laws (Newton), electromagnetism could be understood with a small set of equations (Maxwell), and even linguistics – once a disunified field like neuroscience – now has the basic organizing principles of transformational grammar (Chomsky).

If the brain really is an amalgam of functional circuits built on different, function-specific principles, then we should expect many insights restricted to those circuits, rather than one or just a few truly fundamental insights, as was the case in other scientific disciplines. There is, however, the alternative possibility, one that seems promising to us: that the brain's functional capabilities can be explained by a small but fundamental set of principles that we have not yet recognized, and our ignorance of these principles makes understanding the brain seem much more difficult to us than it really is. Neuroscience has had a long tradition of search for such a set of principles (e.g. James, Sherrington, Pavlov, Hebb, Lashley, Barlow, Edelman). Thus far, this search has failed to achieve its goal. Some even judge this strategy to be bankrupt. We shall address some of their principal objections in the final section of this chapter. But first we will describe the approach we have taken in pursuing the fundamental principles of neuroscience, and present what we believe are very promising results.

2.2.1 Defining the Problem

Where might we look for an insight into the brain's most fundamental principles? It seems to us that those principles might be discerned most readily by considering what makes the brain uniquely special. We suggest that what distinguishes the brain and is at the very roots of its various accomplishments is the ability to make predictions (in the broadest sense).

Prediction and expectation are closely connected. For example, a passing glance at a mostly occluded but familiar object is frequently sufficient to infer its identity, thus incurring expectations for how the object would appear in full view from another angle, or how it would respond to manipulation. Similarly, we can infer the presence of a particular object (e.g. a deer) without actually seeing any part of it, just from circumstantial evidence (e.g. hoof prints). We acquire the expectation that, if we followed the tracks, we would find a deer. In familiar and even not-so-familiar situations we know what will happen next, what to expect. Our own actions are predictive in their very nature: they are carried out in expectation of certain outcomes.

All these are remarkable feats, considering that the brain receives through its senses only very limited and fragmentary information about the outside world. Though the ability to predict comes easily to us, it is computationally difficult: witness the very minimal successes achieved so far by attempts to emulate in artificial systems even the most basic perceptual and motor tasks.

The ability to predict is what makes it possible for humans and other animals to have behaviors successfully adapted to their environments and circumstances. In turn, what enables us to make predictions is the fact that nature is to a large degree orderly. Animals evolved to exploit, via their behavioral interactions with the surroundings, regularities in their environments. The brain is the organ whose *raison d'etre* is to recognize and exploit nature's orderliness. Some regularities are taken advantage of by neural mechanisms that are instinctive. More importantly (especially for advanced animals), other regularities are discovered by the individual, through its sensory experiences and interactions with its surroundings.

What is the mechanism by which the brain discovers regularities? While the job description for this mechanism is simple – the ability to associate related environmental conditions – filling it is far from simple. The main challenge is posed by the fact that the predictable relations among natural phenomena that are most useful behaviorally operate not between pairs of basic environmental conditions, readily detectable by our sensory receptors, but among *combinations* of conditions of various degrees of complexity. To recognize predictable relations among combinations of conditions, the neural mechanism will need to know among which combinations of conditions to look. Unfortunately, there is practically an infinite number of possible combinations to choose from, and most of them do not yield useful regularities.

A common suggestion has been that regularities among environmental conditions are learned by Hebbian connections among neurons participating in representation of these conditions in the brain. The Hebbian rule is an associative synaptic plasticity rule that varies the strength of connections according to temporal correlation in behaviors of the pre- and postsynaptic cells (Hebb, 1949; Brown et al., 1990). However, as stated above, the environmental conditions that are most useful behaviorally are actually complex nonlinear combinations of simple conditions. Is there a mechanism that enables neurons to tune to precisely those combinations of conditions – among the infinite repertoire of possible ones – that are predictable, i.e. associated with other such combinations? This is a very difficult question, for which there is currently no answer (Phillips and Singer, 1997).

In this chapter we propose a radically different mechanism, SINBAD learning, for discovering predictable relations in the environment (SINBAD is an acronym for a *Set of INteracting BAckpropagating Dendrites* referred to in the chapter's title). The virtue of this mechanism is that it does not separate the task of neurons' tuning from that of associative learning; associating *combinations* of environmental conditions and tuning neurons to those *combinations* are two outcomes of the same operation. We will first outline our proposal and then explain the details of its implementation and biological justification.

2.2.2 How Does the Brain Discover Orderly Relations?

We propose that the basic mechanism is implemented at the level of individual neurons. With the cerebral cortex being the part of the brain primarily engaged in the task of discovering regularities, we focus our proposal on the main type of neurons composing the cerebral cortex, pyramidal cells. Pyramidal cells have 5–8 principal dendrites originating from the soma, including 4–7 basal dendrites, and the apical dendrite with its side branches (Feldman, 1984). Each principal dendrite sprouts an elaborate, tree-like pattern of branches and is capable of complex forms of integration of synaptic inputs it receives on its branches from other neurons (Mel, 1994). Through its synapses, each principal dendrite receives information from other neurons about different environmental conditions. Exposed to this information during different situations experienced by the animal, each principal dendrite gradually learns (by adjusting its synaptic connections) to respond in a particular way to patterns of synaptic inputs it receives (Singer, 1995).

Our proposal is that each principal dendrite learns to combine information it receives about different environmental conditions so as to respond to a particular nonlinear combination of them. These combinations are not chosen randomly: each principal dendrite is influenced in its choice by all of the cell's other principal dendrites. Under their mutual influences, principal dendrites in each pyramidal cell choose different, but *co-occurring* or *successive* combinations of environmental conditions. Thus, our basic proposal is that orderly relations in the environment are discovered by individual pyramidal cells, with their principal dendrites each identifying one of several related (and thus mutually predictive) combinations of environmental conditions.

None of the biological learning mechanisms presently known to neuroscience is up to the task of learning nonlinear input-to-output transfer functions (e.g. logical functions, such as *exclusive-OR* or more complex combinations; Phillips and Singer, 1997) and we suggest, as a part of our proposal, a new learning mechanism (which is, in another context, very familiar). It is experimentally well established that synapses on the dendrites of pyramidal cells are capable of modifying their efficacy under the influence of activities of the pre- and postsynaptic cells (Artola et al., 1990; Kirkwood et al., 1993; Markram et al., 1997b). This synaptic plasticity is currently believed to be Hebbian (Singer, 1995). We propose that it resembles Hebbian learning in that it acts in accordance with the Hebbian rule in the context of the experimental conditions under which it has been studied, but that it is actually a more complex form of learning. According to the form of synaptic plasticity that we propose, the strength of a synapse is controlled not only by the two factors that are used in Hebbian plasticity (output activities of the pre- and postsynaptic cells), but also by the output activity of the principal dendrite on which that synapse resides. Synaptic strength is controlled on the postsynaptic side by the *difference* between the output activity of the postsynaptic cell and the output activity of the host principal dendrite. Such a form of learning is known in the artificial neural network literature as *error-correction* learning (Widrow and Hoff, 1960; Rumelhart et al., 1986). The postsynaptic controlling factor – the difference between the cell's and the dendrite's outputs – is traditionally called the *error*, or *delta*, signal and the effect of this type of learning is to minimize the error signal (Widrow and Hoff, 1960).

As a part of synaptic strength modification, the "error" signal is propagated back to each synapse. If local dendritic branches integrate their inputs nonlinearly (Mel, 1994), then to be effective, the error signal delivered to each synapse will have to be modified in a certain way according to the dendritic conditions along the path from the synapse to the soma. The nature of this modification is well understood theoretically and is known as *error backpropagation* (Rumelhart et al., 1986). Thus we propose that dendrites of pyramidal cells learn by "error" backpropagation; in other words, a principal dendrite is a form of the well-known backpropagation ("back-prop") network.

In standard applications of backpropagation networks, the desired output patterns, used to train the network, are provided by an external "teacher". In the dendritic application of backpropagation learning, the role of such a teacher is played by other dendrites: each principal dendrite is "taught" by all the other principal dendrites of the same cell. The purpose of this arrangement is for each principal dendrite in a given pyramidal cell to learn to predict, on the basis of the synaptically-transmitted information available to *it*, what the cell's *other* principal dendrites are doing in response to *their* inputs. The functional significance of the dendrites learning to predict (i.e., match) each other's activities is that the dendrites will identify different combinations of environmental conditions that are predictive of each other. The cell, as a whole, tunes to a set of related combinations of environmental conditions, which define an orderly feature of the environment, such as an object, a complex property, a causal relation, etc.

Combinations of environmental conditions learned by principal dendrites have a number of useful properties. They are *predictable*: that is how they are identified in the first place. They are *informative*: being involved in orderly relations, they are predictive of other conditions taking part in those relations. Finally, they can be useful as *building blocks* in the construction – by other pyramidal cells – of higher-order combinations of environmental conditions and the discovery of regularities involving those higher-order conditions. The path to discovery by the cerebral cortex of high-order regularities in the environment is through discoveries of lower-order regularities, from simple to progressively more complex, because in nature higher-order regularities are built from lower-order regularities. Because lower-order regularities are simpler, they will be easier for cortical pyramidal cells to recognize. In turn, the recognized regularities will make it easier for pyramidal cells in higher-level cortical areas to recognize higher-order regularities that involve them. The recognized higher-order regularities will enable recognition of regularities of even higher order (and in addition via feedback connections they might help recognize other lower-order regularities as well), and so on. Thus, for example, recognition of lines, edges, and textures by some pyramidal cells will enable recognition of surfaces and figures by other cells, which in turn will enable recognition of different types of objects and their states by yet other cells, which will enable recognition of different types of situations (involving interacting objects), etc.

In conclusion, we propose that pyramidal cells of the cerebral cortex are devices for discovery of orderly features of the environment. An individual pyramidal cell obviously has a limited ability to recognize orderly features in the information it receives from other cells. But, when organized into sequences of cortical networks, pyramidal cells can build their discoveries on the discoveries of the preceding cells, thus gradually unraveling nature's more and more complex relations.

2.3   Implementation of the Proposal

2.3.1 How Might Error Backpropagation Learning Be Implemented in Dendrites?

Our proposed mechanism for discovering orderly environmental properties is built on dendrites of pyramidal cells learning by error backpropagation. Figure 2.1 illustrates how a principal dendrite of a pyramidal cell can be treated as a backpropagation network. Figure 2.1A shows a drawing of an idealized principal dendrite emanating from a pyramidal cell soma. This dendrite has one primary, two secondary, and four tertiary dendritic branches. Locations of synaptic contacts on those branches are also shown. The individual dendritic branches can be thought of as separate compartments (i.e. electric circuits; Segev et al., 1989), and the entire dendritic tree can be thought of as a system of such interconnected compartments (Figure 2.1B). Each compartment receives direct synaptic contacts from other cells, and it is also connected with other, contiguous compartments.

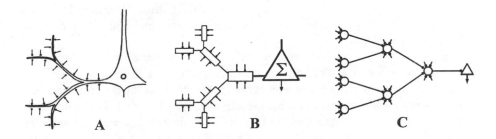

Figure 2.1. A principal dendrite of a pyramidal cell (A) viewed as a set of connected compartments (B) and as a multilayered backpropagation network (C). Also shown are synaptic connections, distributed throughout the dendritic tree.

The final step in the transformation from the anatomical view of principal dendrites to their representation as a backpropagation network is shown in Figure 2.1C. As shown here, each dendritic compartment can be treated as a hidden unit of a backpropagation network. Each compartment receives "input layer" (not drawn explicitly) connections (i.e., synaptic connections) and, if it is the primary or a secondary dendritic compartment, it also receives connections from the preceding "hidden layer" (i.e., from more distal compartments).

The key proposal here is to treat the link between two dendritic compartments as an activity-modifiable connection. In physiological terms, the strength of the connection between two compartments is measured as a conductance, and it can be modified in a number of ways, including changes in passive membrane conductance or local changes in the dendritic diameter. The last mechanism is quite interesting because it would predict that learning involves not only changes in synaptic efficacy, but also changes in the sizes and the morphology of dendrites, possibly including retractions as well as sproutings of local dendritic branches (Quartz and Sejnowski, 1997).

In accordance with the backpropagation design, dendritic compartments integrate their inputs nonlinearly, a biologically realistic proposition (Mel, 1994). Dendritic compartments should integrate their inputs nonlinearly because this will enable principal dendrites to learn more complex, linearly inseparable combinations of environmental conditions and will give the dendrites incomparably greater information-processing powers. Some variations in nonlinear properties in different parts of dendritic trees are acceptable: e.g. while distal dendritic compartments should integrate inputs nonlinearly, the proximal compartments can be linear.

The primary dendritic compartment plays the role of the output unit of the backpropagation network; its output is compared with the somal output (which plays the role of a "teacher" signal). The primary compartment, being the stem of the entire dendritic tree, is in an advantageous position to compute the difference between the outputs of the soma and the principal dendrite. First, all the synaptic inputs onto the entire tree travel and converge at the stem, where they summate and thus compute the output of the entire tree. And second, coming from the opposite direction, from the soma, the stem of the dendritic tree also receives information about the cell's output in the form of action potentials. We propose that the computed difference, or the error signal, is then back propagated to all the more distal dendritic compartments and is used to adjust local connections, both the synaptic connections and the connections between compartments.

For demonstration purposes we modeled error backpropagation in the pyramidal cell shown in Figure 2.1. The cell was given external inputs from four other cells. Each input cell made connections with every compartment of the dendritic tree (it is not necessary for the success of backpropagation learning that each input cell should connect to every compartment, but in this first, simple example we have very few input cells and very few compartments). These connections were controlled by the backpropagation algorithm. The primary and secondary dendritic compartments also had connections – again controlled by the backpropagation algorithm – from two more distal compartments, as shown in Figure 2.1C.

The activity of each tertiary dendritic compartment was computed as a sigmoid function of its inputs (in this case, hyperbolic tangent with a range between -1 and 1)

$$A_i = \tanh(\sum_{j=1}^{4} w_{ji} \cdot IN_j), \tag{1}$$

where A_i is the activity of compartment i, j is an input cell and IN_j and w_{ji} are its activity and its connection strength to compartment i. Activities of the secondary dendritic compartments, which had additional connections from tertiary compartments, were computed analogously

$$A_i = \tanh(\sum_{j=1}^{4} w_{ji} \cdot IN_j + \sum_{k=1}^{2} w_{ki} \cdot A_k), \tag{2}$$

where k is a compartment distal to compartment i, and A_k and w_{ki} are its activity and its connection strength to compartment i.

The activity of the primary dendritic compartment was computed as a sum of its inputs

$$A_i = \sum_{j=1}^{4} w_{ji} \cdot IN_j + \sum_{k=1}^{2} w_{ki} \cdot A_k), \tag{3}$$

where k is a secondary compartment, and A_k and w_{ki} are its activity and its connection strength to the primary compartment.

The activity of the primary compartment is also the output of the entire principal dendrite. It contributes to the output of the entire pyramidal cell, i.e. the somal output

$$OUT = TR + A_1. \tag{4}$$

Here, TR is the training signal, representing the net contribution to the somal output from all of the cell's *other* principal dendrites.

The associative learning task we chose for this demonstration is learning a logical function that cannot be learned by a linear neural network (because the training stimulus patterns are not linearly separable), but is easily learned by a standard nonlinear backpropagation network. The function we chose is:

$$TR = (IN_1 \quad XOR \quad IN_2) \quad \& \quad (IN_3 \quad XOR \quad IN_4). \tag{5}$$

This function describes a relationship (& and XOR are the logical functions *AND* and *exclusive-OR*, respectively) between the training signal TR and the pattern of activities of the four input cells $IN_1 - IN_4$. The entire training set of 16 stimulus patterns is shown in Figure 2.2.

After initially setting all the adjustable connections to randomly chosen strengths w's, the cell was activated with a randomly chosen sequence of 16 training stimulus patterns. The connections were adjusted according to the error backpropagation algorithm (Rumelhart et al., 1986) after each stimulus presentation. Specifically, the error signal δ was first computed for the primary dendritic compartment as

$$\delta = OUT - \alpha \cdot A_1, \tag{6}$$

where α is a scaling constant that has to be greater than 1 in order for the dendrite's output to have a net negative contribution to the error signal; in our demonstration $\alpha = 2$.

For the secondary and tertiary basal compartments δ was computed as

$$\delta_i = \delta_j \cdot w_{ij} \cdot A_i', \tag{7}$$

where i is the compartment for which δ is computed, and j is the more proximal compartment to which compartment i connects. Connection strengths were adjusted by

$$w_{ij}(t+1) = w_{ij}(t) + \mu \cdot A_i \cdot \delta_j, \qquad (8)$$

where μ is learning rate constant ($\mu = 0.1$ in our simulations), i is a more distal dendritic compartment or an input cell (in which case $A_i = IN_i$), and j is a compartment that receives the connection.

It would seem to be a departure from biological realism that this connection-adjusting algorithm (Equations 6–8) can allow a change in the sign of a connection (or, in physiological terms, it can allow a change from excitatory to inhibitory and vice versa). With regard to synaptic connections, for biological realism our setup should be interpreted as follows: each input cell in the model actually stands for two input cells that have identical activities but different physiological sign – one is excitatory, the other is inhibitory (for experimental evidence of plasticity of inhibitory connections in the cortex, see Komatsu and Iwakiri, 1993; Komatsu, 1994; Pelletier and Hablitz, 1996). These cells cannot change the sign of their connections; for example, when a connection weight of an excitatory cell should, according to Equation 8, go below zero, it is just set to zero. At the same time, the connection of the inhibitory counterpart cell – which until now has been set to zero – now acquires a negative (i.e. inhibitory) value. Thus, a connection weight w in the model actually describes the weights of two functionally identical, excitatory and inhibitory input cells.

With regard to connections between dendritic compartments, their weights should only be positive, since they represent positive longitudinal conductances along the dendrites. Accordingly, in the model the weights of intercompartmental connections were never reduced below a certain minimal positive value, which in the case of this demonstration was set at 0.01.

Training of the connections (the training stimuli were presented in a random sequence) continued until their strengths came close to stable values. At this time we evaluated the performance of the dendritic tree by presenting all 16 training stimulus patterns and observing the dendritic output A_i. Results are plotted in Figure 2.2, showing that the dendrite successfully learned to accurately predict the training signal: when $TR = 1$, A_i was very close to 1, and when $TR = 0$, A_i was very close to zero.

This exercise demonstrates that dendrites of pyramidal cells can, in principle, implement the error backpropagation learning algorithm, enabling them to learn complex, linearly inseparable functions, and to capture complex relations that are beyond the reach of the Hebbian rule. What is not resolved by this demonstration, however, is whether dendrites do in fact learn by error backpropagation. The current understanding, based on extensive experimental work (Singer, 1995), is that synaptic learning in the cerebral cortex is controlled by the Hebbian rule, according to which the strength of a synapse is a function of the correlation in behavior of the pre- and postsynaptic cells. However, this experimental work does not preclude the possibility that the synaptic learning rule in the cortex is the error-correcting one: the two rules, while greatly different in their functional consequences, are quite similar in their details. Experimentally the error-correction rule can easily resemble the Hebbian rule. In fact, the error-correction rule can be described as an elaboration of the Hebbian rule by an additional controlling factor, namely, the local dendritic activity along the path from the synapse down the dendritic tree to the soma.

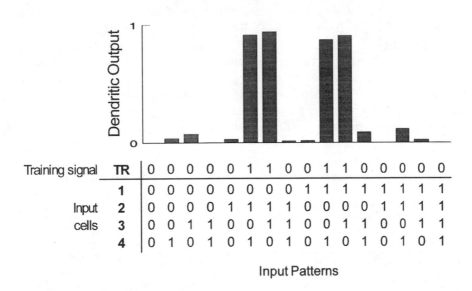

Training signal	TR	0	0	0	0	0	1	1	0	0	1	1	0	0	0	0	0
	1	0	0	0	0	0	0	0	0	1	1	1	1	1	1	1	1
Input	2	0	0	0	0	1	1	1	1	0	0	0	0	1	1	1	1
cells	3	0	0	1	1	0	0	1	1	0	0	1	1	0	0	1	1
	4	0	1	0	1	0	1	0	1	0	1	0	1	0	1	0	1

Input Patterns

Figure 2.2. Output of the principal dendrite, A_1, in response to the 16 training stimulus patterns. The training patterns are shown in the table above, aligned with the plot.

Whether it is Hebbian or error-correcting, the synaptic learning rule requires that a signal describing the output activity of the postsynaptic cell be backpropagated from the axon hillock (where output action potentials originate) to each individual synapse. The most likely means by which information about a cell's output reaches the sites of individual synapses on distal dendrites is via spikes that are actively back propagated from the soma up the dendrites (Markram et al., 1997b; Stuart et al., 1997). In the Hebbian rule, the postsynaptic activity signal delivered to a synaptic site should reflect accurately and without distortions the cell's output activity. In the error-correction rule this signal is modified in two ways. The first is by subtracting the activity of the primary dendritic compartment (i.e. the most proximal portion of the dendrite) from the cell's output activity (see Equation 6). Thus the error-correction learning rule can be seen as a version of the Hebbian rule where each principal dendrite's output activity attenuates the postsynaptic signal backpropagated through it more than it contributes to it. The second modification of the postsynaptic signal is due to its re-scaling by local longitudinal conductances (probably reflecting diameters of dendritic branches) and by local activity along the dendritic path from the soma to the synapse (see Equation 7). Both modifications are physiologically very

plausible, and experimental and theoretical evidence suggests that some of such modifications of the postsynaptic signal do take place (Stuart et al., 1997).

In conclusion, there are grounds to be optimistic that synaptic learning in dendrites of cortical pyramidal cells is of the error correction type, implemented by error
backpropagation in the dendrites. To evaluate this possibility, we need experiments in
which the efficacy of synaptic connections is studied as a function of not only the
pre- and postsynaptic output activities, but also the activity of the host dendrite. The
actual learning rule used by the dendrites probably will deviate in its details from the
one we describe here (Equations 6–8). It might be more effective than the modern
backpropagation algorithms used in artificial neural networks. On the other hand, it
might be less effective, since learning demands placed on individual dendrites are
likely to be quite modest, and the learning power of the entire cortex arises from having large numbers of cells, and from having multiple hierarchically organized cortical
areas. What is required from the type of learning we propose here is that each dendrite is capable of tuning to *nonlinear* combinations of its inputs that will *minimize*
the difference between the output of that dendrite and the output of the entire cell.

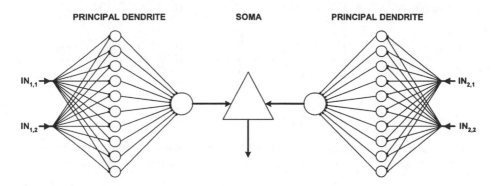

Figure 2.3. The SINBAD model of a pyramidal cell with two principal dendrites connected to
the soma (shown as a triangle). Each principal dendrite is modeled as an error backpropagation
network with one output unit, a single layer of ten hidden units, and two input channels.

2.3.2   How Can Dendrites Be Set Up to Teach Each Other?

Interpreting the principal dendrites of cortical pyramidal cells as functional equiva-
lents of backpropagation networks is the first part of our proposal. Backpropagation
learning is a supervised form of learning, and the second part of the proposal is that
the role of a teacher for a principal dendrite is played by the other principal dendrites
of the cell. To explain how dendrites can teach each other and what they can learn in
the process, we set up a model of a single pyramidal cell with two principal dendrites.
Real pyramidal cells have 5–8 principal dendrites, but two dendrites are sufficient for
presenting our basic idea.

Each principal dendrite is modeled in this demonstration as a standard backpropagation network, rather than as a dendritic tree, shown in Figure 2.1. The reason for
this choice is that in our view the principal dendrite is essentially a functional equiva-

lent of the backpropagation network, and the backpropagation network is easier to model and understand. The design of the cell is shown in Figure 2.3. Each dendrite consists of 10 hidden units and one output unit representing the primary compartment of the dendrite. The hidden units of each dendrite receive input connections from two other cells, or *input channels*, that carry information about environmental conditions. In the present demonstration the two principal dendrites receive connections from two different pairs of input channels, $IN_{1,1}$ and $IN_{1,2}$, versus $IN_{2,1}$ and $IN_{2,2}$.

Activity of a hidden unit h in dendrite d is computed as a sigmoid function of activities of its two input channels

$$H_{d,h} = \tanh(w_{d,1,h} \cdot IN_{d,1} + w_{d,2,h} \cdot IN_{d,2}), \tag{9}$$

where $w_{d,1,h}$ and $w_{d,1,h}$ are the weights of connections of two input channels $d,1$ and $d,2$ on hidden unit h of dendrite d.

Activity of the output unit, i.e. the output of dendrite d, is:

$$D_d = \sum_{h=1}^{10} w_{d,h} \cdot H_{d,h}, \tag{10}$$

where $w_{d,h}$ is the weight of the connection of hidden unit d, h to the output unit.

Outputs of the two dendrites are summated to produce the somal input $SOM = D_1 + D_2$. The output of the entire cell is

$$OUT = \tanh(\gamma \cdot SOM), \tag{11}$$

where γ is a variable that adjusts the cell's output by whether somal input is greater or smaller than the average somal input. Specifically, $\gamma = 1.2$ if $|SOM| > \overline{|SOM|}$ and $\gamma = 0.8$ if $|SOM| \le \overline{|SOM|}$. This adjustment drives the cell to expand the dynamic range of its output values. For demonstrating our idea, it is not important here that, in deviation from biological realism, the cell's output can be either positive or negative.

The sensory environment of the cell was conceived to involve a single orderly entity, object X, that can be either present or absent in any given situation (i.e. $X = 1$ or $X = 0$). Object X manifests itself in two combinations of environmental conditions, each of which is represented by the activity (0 or 1) of one of the input channels. Specifically,

$$X = IN_{1,1} \ \textit{exclusive-OR} \ IN_{1,2} , \tag{12}$$

and

$$X = IN_{2,1} \ \textit{exclusive-OR} \ IN_{2,2} . \tag{13}$$

That is, when present, object X can reveal itself by activating either input channel $IN_{1,1}$ or $IN_{1,2}$, but not both of them, and also by activating either channel $IN_{2,1}$ or $IN_{2,2}$, but not both of them.

There are eight possible patterns of input channel activities that satisfy Equations 12 and 13, and they define the entire repertoire of environmental situations distinguished by the input channels. All these patterns are shown in Figure 2.4B. An inspection of these patterns will reveal that the sensory environment is orderly, but this order is not reflected in activities of single channels, and is only apparent at the level of specific (*exclusive-OR*) combinations of two pairs of channels. Can the cell discover that the combining functions defined by Equations 12 and 13 are predictive of each other and also informative of their underlying causal source, object *X*? Note that the existence of object *X* is not indicated to the cell in any direct way, but it is only hinted at indirectly by the regularities hidden in the input patterns.

After initially setting all the adjustable connections to randomly chosen strengths w's, the cell was activated with a randomly chosen sequence of the eight training input patterns. The connections were adjusted according to the error backpropagation algorithm after each input pattern presentation. Specifically, the error signals δ_d were first computed for the two dendrites as

$$\delta_d = (OUT - EST_d) \cdot EST'_d = (OUT - EST_d) \cdot (1 - EST_d^2), \quad (14)$$

where EST_d is the dendrite d's estimate of the cell's output OUT. It was computed as

$$EST_d = \tanh(2 \cdot D_d). \quad (15)$$

For the hidden units, δ was backpropagated as

$$\delta_{d,h} = \delta_d \cdot w_{d,h} \cdot H'_{d,h}. \quad (16)$$

Connection weights were adjusted by

$$\Delta w_{d,i,h} = \mu_i \cdot IN_{d,i} \cdot \delta_{d,h} \quad (17)$$

and

$$\Delta w_{d,h} = \mu_h \cdot H_{d,h} \cdot \delta_d, \quad (18)$$

where μ_i and μ_h are learning rate constants for the input and hidden unit connections ($\mu_i = 6$ and $\mu_h = 0.003$ in our simulations).

Before we turn to the results of simulation of this model neuron, it might be helpful to consider the nature and dynamics of the learning task we set up for the two dendrites. Being a backpropagation network, each dendrite can, in principle, learn a large variety of input-to-output transfer functions. Its actual choice will be dictated by the teaching signal, coming from the other dendrite. But that dendrite also has a large choice of possible transfer functions, and it relies on the first dendrite for its own guidance. Thus, the two dendrites will teach each other how to respond to input patterns while continuously changing their own behaviors and their own teaching signals. Such a teaching/learning process will continue until the two dendrites discover such transfer functions that will enable them to have identical responses to their *different* co-present inputs. In other words, the process of connection strength adjustments will continue until each dendrite will learn to predict – on the basis of its own inputs – the responses of the other dendrite to its inputs.

Of course, if the input patterns applied to one dendrite do not relate in any way to the input patterns applied at the same time to the other dendrite, but are accidental in their co-occurrence, then it will be impossible for the dendrites to discover any matching transfer functions, and the process of connection strength adjustments will continue indefinitely. On the other hand, if there is some consistent, and therefore predictable, relationship between co-occurring patterns of the input to the two dendrites, then the dendrites might be able to discover transfer functions predictive of each other's outputs. Success will not be guaranteed, but will depend on the complexity of the orderly relations between the two sets of inputs. For our demonstration here we chose an intermediate level of complexity of the orderly relationship between input channels $IN_{1,1} - IN_{1,2}$ and $IN_{2,1} - IN_{2,2}$; as described above (Equations 12–13), this relationship is not discernable at the level of individual channels, but only at the level of their *exclusive-OR* logical combinations.

Figure 2.4 shows the progress and the results of the two dendrites teaching each other how to respond to input patterns. The most pressing question is: Will the two dendrites learn to predict each other's responses to the input patterns? To answer this question, the difference in the output activities of the two dendrites (i.e. $|D_1 - D_2|$) was plotted in Figure 2.4A as a function of each successive input pattern presentation. This plot shows that the two dendrites discovered very quickly how to respond to their inputs so that they will produce identical outputs. Exposure to a random sequence of fewer than 20 input patterns was sufficient for the two dendrites to discover a way to produce nearly identical responses to their different inputs.

The dendrites' learning success raises the following question: What did the dendrites discover about the environment that enabled them to predict each other's responses? To answer this question, we need to examine responses of the whole cell to the entire repertoire of eight possible input patterns. These responses (plotted in Figure 2.4B) were obtained after the cell was exposed to a random sequence of 100 input patterns, by which time the two dendrites already learned to respond virtually identically (see Figure 2.4A).

Judging by the cell's responses plotted in Figure 2.4B, the first dendrite learned to respond to the exclusive-OR combination of its input channels $IN_{1,1} - IN_{1,2}$, while the second dendrite learned to respond to the exclusive-OR combination of its input channels $IN_{2,1} - IN_{2,2}$. Because of the way we set up the sensory environment (Equations 12–13), this pair of transfer functions is the only pair that produces identical outputs, and that is why the dendrites chose them. Thus, Figure 2.4B shows that the two dendrites successfully identified two nonlinear combinations of their inputs that have an orderly, predictable relationship.

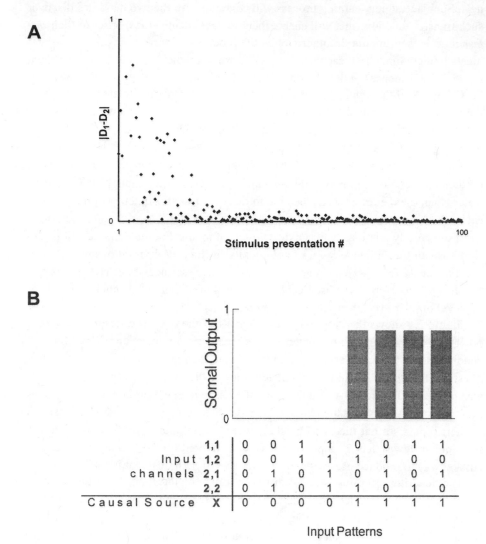

Figure 2.4. Learning performance of the SINBAD model. (A) Magnitude of the difference in outputs of the two dendrites in response to a random sequence of input patterns. In response to each input pattern, the two dendrites adjusted their connections, showing a rapid learning progress, which was essentially completed by the 20th stimulus presentation. (B) Output of the cell, *OUT*, in response to the eight training input patterns. The training patterns are shown in the table, aligned with the plot. Note the match between the state of the causal source X and the cell's output.

The functional significance of the outcome of the dendrites' learning goes beyond the discovery of two orderly combinations of environmental conditions; it identified the causal source of this order. This causal source is object X: the reason why the two combinations of conditions are predictive of each other is because they originate from the same source, object X. As Figure 2.4B shows, the cell's output is accurately indicative of the presence and absence of object X. The cell, in effect, discovered the existence in the environment of object X.

In conclusion, this modeling exercise shows that by teaching each other, the dendrites in a cell can tune to different combinations of environmental conditions that are predictive of each other and the cell, as a whole, can learn to recognize orderly features in its sensory environment.

One necessary condition for cells to be able to discover orderly environmental features is that different dendrites on the same cell should receive connections from different input cells. If all the dendrites in a cell are exposed to the same input sources, then they will not need to discover predictable relations in the environment: having the same inputs, they can easily agree to produce the same outputs, whatever they happen to be. It is only when the dendrites have different input sources that they are forced to rely on regularities in the environment in order to find something common in their different inputs. Thus, for example, if each dendrite in our model had connections from all four input channels, the dendrites would learn transfer functions that produced identical outputs, but most likely they would have no relationship to object X; i.e. the dendrites would fail to discover the two combinations of the input channels that are orderly.

There are multiple reasons to believe that real pyramidal cells in the cerebral cortex do indeed satisfy this condition and have their principal dendrites exposed to different input sources. First, it is very unlikely that any given axon making a synaptic contact on one dendrite will also have synapses on *all* the other principal dendrites of the same cell (Schuz, 1992; Thomson and Deuchars, 1994). It can, and frequently does, have synapses on more than one principal dendrite, but not on all of them (Deuchars et al., 1994; Markram et al., 1997a).

Second, connections coming from different sources have prominent tendencies to terminate on different dendrites. For example, neighboring pyramidal cells make synaptic connections preferentially on the basal dendrites (Markram et al., 1997a), whereas more distant cells, including ones several millimeters away in the same cortical area, terminate preferentially on the apical dendrites (McGuire et al., 1991). Another system of connections, the feedback connections from higher cortical areas, terminate preferentially on yet another part of the dendritic tree, i.e. on the terminal tuft of the apical dendrite, located in layer I (see Larry Cauller's Chapter 1 in this volume). This terminal tuft, although not originating directly from the soma, might functionally be considered a principal dendrite, due to its special means of communication with the soma.

The third source of differences in inputs to different principal dendrites of the same cell is cortical topographic organization. Across a cortical area, functional properties of cells change very quickly: even adjacent neurons carry in common less than 20% of stimulus-related information (Gawne et al., 1996; for a review, see Favorov and Kelly, 1996). Basal principal dendrites extend in all directions away from the soma and thus spread into functionally different cortical domains. As a result, these dendrites sample different loci of the cortical topographic map and are exposed to different aspects of the cortical representation of the environment (Malach, 1994).

We elaborate this idea further in Section 2.4, which provides a brief review of the fine, minicolumnar structure of cortical topographic maps and relates the SINBAD proposal advanced in this chapter to the cortical minicolumnar organization.

Overall, the 5–8 principal dendrites of the same pyramidal cell are exposed to different sources of information about the environment and, in order to have correlated output behaviors, they will have to tune to different but mutually predictive combinations of environmental conditions.

2.3.3   How to Divide Connections Among the Dendrites?

So far, we have described the principle by which pyramidal cells can discover regularities in their sensory environment. For any practical implementation of this principle we must address the question of how to distribute input channels among the principal dendrites of a cell so that the right combinations of channels will go to the right dendrites. In the previous modeling exercise we skirted this issue by assigning the four input channels to the two dendrites in pairs that we knew were the right ones for the environment we set up there. But if we did not know the orderly organization of the environment in advance, then we would not know how to divide the four channels between the two dendrites. Another concern is that, unlike in the previous exercise, the environment possesses not just one, but many different orderly features and of various levels of complexity. How can the cells discover as many of these orderly features as possible?

To address these issues, we will start with another simple modeling demonstration. For this demonstration we created a more complex sensory environment, which involved regularities of three orders of complexity. This environment is characterized by 32 parameters, or *elementary conditions*. They might be, for example, 32 sensory channels through which some hypothetical animal obtains information about the state of its surroundings. The environment is orderly, which means that only a limited subset of all possible combinations of elementary conditions can occur. Figure 2.5 shows a number of examples of such orderly patterns, and these examples make it obvious that the orderly features of this environment will not be easy to recognize.

To appreciate the difficulty in making use of the orderly structure of this environment, we can consider a hypothetical animal that inhabits it. Suppose a particular third-order combination of environmental conditions, involving some of the environment's orderly features, has a specific behavioral significance to an animal. In Figure 2.5, a small sample of the 2^{22} possible environmental states are separated into two panels according to the presence or absence of that combination. Consider the two panels. How to distinguish between these two sets of patterns? The approach advocated in this chapter is to start by learning to recognize first-order regular combinations among elementary conditions, then second-order regular combinations among the first-order ones, and finally the third-order combination.

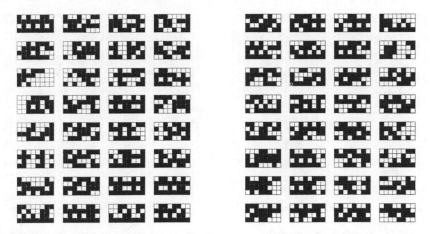

Figure 2.5. Sixty-four examples (out of 2^{22} possible) of orderly states that can be taken by the model environment. Each state is shown by an 8×4 field of small squares, with each square representing (by its shading) the status of one of the 32 elementary environmental conditions (white = absence, or 0; black = presence, or 1). The environmental states shown in the two panels can be distinguished by the presence (right panel) or absence (left panel) of a particular third-order combination of elementary environmental conditions, described in the text.

The orderly structure of the modeled environment is illustrated in Figure 2.6. The environment was set up to have two entities, called objects α and β. The animal should act one way when either object α or object β is present, and it should act the other way when neither or both objects are present (this is an *exclusive-OR* logical function).

Object α in turn, is indicated by environmental properties a and b, let's say α = *(a exclusive-OR b)*. That is, object α manifests itself as a or b, whereas together a and b do not indicate α, but form an accidental combination.

In addition, α can be reliably predicted by a combination of properties c and d, let's say α = *(c exclusive-OR d)*. It might be, for example, that α has two sides, one side characterized by either a or b and the other by either c or d combinations. Or the cd combination might describe some other environmental condition that co-occurs with object α.

Object β is organized according to the same plan as object α: β = *(e exclusive-OR f)*, and β also can be reliably predicted by a combination *(g exclusive-OR h)*.

To add one more layer of complexity, environmental properties a, b, c, d, e, f, g, h are in turn defined by elementary environmental conditions *1* through *32*. All of these properties were given the same organization:

a = *(1 exclusive-OR 2)* = *(3 AND 4)* b = *(5 exclusive-OR 6)* = *(7 AND 8)*
c = *(9 exclusive-OR 10)* = *(11 AND 12)* d = *(13 exclusive-OR 14)* = *(15 AND 16)*
e = *(17 exclusive-OR 18)* = *(19 AND 20)* f = *(21 exclusive-OR 22)* = *(23 AND 24)*
g = *(25 exclusive-OR 26)* = *(27 AND 28)* h = *(29 exclusive-OR 30)* = *(31 AND 32)*

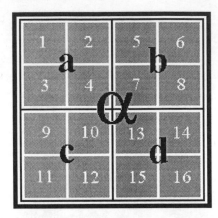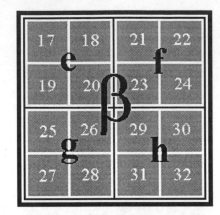

Figure 2.6. The orderly structure of the model environment. Shown as small gray squares are 32 elementary environmental conditions, organized into eight sets that define environmental properties *a–h*. These eight properties, in turn, are organized into two sets that define objects α and β. The rules by which lower-order conditions define the status of higher-order conditions are specified in the text.

Thus, the modeled environment possesses 32 elementary conditions, eight first-order predictable combinations of elementary conditions (*a–h*), and two second-order predictable combinations (α, β).

We presented this environment to the model used in the previous demonstration, with a few modifications. The first modification is that this cell now has 32 input channels, each carrying information about one of the 32 elementary environmental conditions. Unlike the previous exercise, these connections are divided randomly between the two dendrites of the cell. Each input channel is assigned at random to either one or the other dendrite, but not to both of them. Also, connection weights of input channels are set initially to zero, except for four randomly chosen channels on each dendrite. For these four connections, their initial connection weights are chosen at random. Thus, unlike the previous exercise, here we do not take advantage of our knowledge of which input channels should go together.

In another modification of the initial design, the number of hidden units in each dendrite is increased to 40. Otherwise, activities of the hidden units $H_{d,h}$ and the dendrite's output unit D_d are computed as described in Equations 9 and 10. The cell's output *OUT*, error signal δ_d, error backpropagation $\delta_{d,h}$, and dendritic connection *w* adjustments are computed as before, according to Equations 11 and 14–18.

Figure 2.7 shows the result of one simulation run during which 10,000 input patterns were presented in a random sequence to the cell. To see whether the two dendrites of the cell learned to produce similar outputs in response to their co-present input patterns, the correlation coefficient between outputs of the two dendrites was computed across 100 successive input pattern presentations. In Figure 2.7 the value of this correlation coefficient is plotted as a function of time since the start of the training period. This plot shows that in the beginning the two dendrites correlated poorly in their outputs, but after some initially unsuccessful search the dendrites discovered a way to produce very similar outputs. This means that the two dendrites discovered some orderly relationship in the environment, which enabled them to predict each other's responses to input patterns.

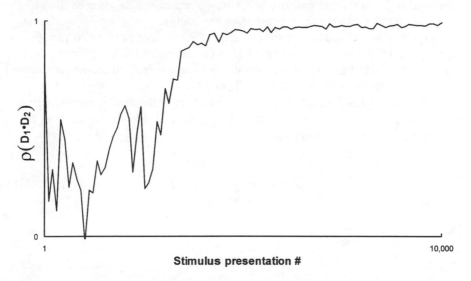

Figure 2.7. Learning performance of the SINBAD model. The correlation coefficient, ρ, between the outputs of the two dendrites is plotted as a function of time from the start of the training period.

The environment has a number of orderly features, listed above, and which of them were discovered by the cell was determined by the distribution of the input channels on the two dendrites. Because the channels were distributed between the dendrites randomly, there is no guarantee that a set of channels engaged in a given orderly combination (e.g. channels 1 and 2, as an exclusive-OR combination, reporting on the status of property a were placed on the same dendrite, or that predictive pairs of channels were placed on the opposite dendrites (e.g. channels 1 and 2 on one dendrite and channels 3 and 4 on the other dendrite). As a result, the particular distribution of input channels on the dendrites might make it impossible for the cell to discover a particular environmental regularity.

There are a number of ways a cortical network is likely to deal with this limitation. The simplest way is to have many pyramidal cells. Some of them will, by accident, have an appropriate distribution of input channels on their dendrites and thus they will be in a position to discover that orderly environmental feature. With a sufficient number of cells, the network will be able to discover all the orderly features.

To illustrate this idea, we expanded the model to have 32 cells identical to the one used in the last exercise. They differ from each other only in the distribution of input channels on their dendrites, which was assigned randomly. For simplicity, the 32 cells were run in parallel, without interactions among them. To evaluate the learning outcomes of this network, we also set up three additional cells (or more accurately one of the principal dendrites of three additional cells residing in a higher cortical area). The design is shown in Figure 2.8. Our aim here is to compare the ability of a "test" dendrite to learn a particular behavior, given either raw information, provided by the input channels, or the information provided by the 32 cells (which presumably transformed the raw input patterns into a new form in which some of the orderly environmental features are represented more explicitly).

One test dendrite was trained to respond to the presence of object α, another to object β, and the third to a combination $\Omega = (\alpha$ *exclusive-OR* $\beta)$. (These dendrites might be viewed as belonging to cells that, for some unspecified reasons, are driven by their other dendrites to respond to α, β, or Ω.)

The three test dendrites are modeled the same way as the dendrites of the 32 cells, except that the test dendrite receives connections from all 32 cells (or 32 input channels, in the raw input test) and its output is passed through a sigmoid function:

$$D = \tanh(\sum_{h=1}^{40} w_h \cdot H_h). \tag{19}$$

The error signal is computed as

$$\delta = (TR - D) \cdot D', \tag{20}$$

where TR is the training signal (the status of α, β, or Ω). Effectively, these test dendrites are modeled as standard three-layer backpropagation networks.

To establish the basis for comparison, we first trained the test dendrites on raw sensory information, by using input channels $IN_1 - IN_{32}$ for their inputs. The results are shown in Figure 2.9A, for the test dendrite trained on object α (see the plot on the left) and for the test dendrite trained on combination α (right plot). Each plot shows the error (which is computed as $|TR - D|$) as a function of time since the start of the training period. As the left plot shows, the α dendrite gradually succeeded in learning to recognize object α. The right plot, on the other hand, shows that the Ω dendrite failed to learn to recognize combination Ω. This failure is not surprising, since Ω is a third-order combination, and backpropagation networks have great difficulties learning such complex functions (Clark and Thornton, 1997).

We next trained the test dendrites on sensory information processed by the 32 cells: outputs of the 32 cells were used as inputs to the test dendrites. The results are shown in Figure 2.9B, again for the α dendrite on the left and for the Ω dendrite on the right. As these two plots show, the learning performance of the test dendrites improved dramatically. The α dendrite learned to recognize object α orders of magnitude faster than when it learned from the raw input channels. And the Ω dendrite – which before could not learn at all – now did learn to recognize combination Ω. What is particularly impressive is that the test dendrites learned to respond correctly to the input patterns after being presented with only a small fraction (<0.5%) of all possible input patterns. That is, the test dendrites showed perfect generalization abilities; they discovered the logic underlying the relationship between the input patterns and the behaviors on which they were trained.

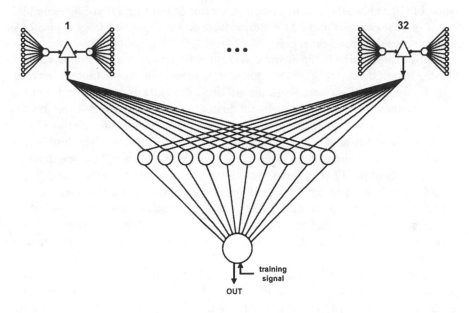

Figure 2.8. A layer of 32 SINBAD cells whose outputs are connected to a test dendrite, which is modeled as a backpropagation network with one output unit and a single layer of hidden units.

The performance of the Ω dendrite is especially impressive as it was able to learn a very challenging, third-order combination of input channels. This task seemed impossible when we discussed it earlier in the example of a hypothetical animal that needed to learn how to distinguish between the two types of input patterns shown in Figure 2.5. After training, the Ω dendrite became able to distinguish between them; it happens (by our design) that the two sets of patterns in Figure 2.5 differ in that in one set the Ω combination of environmental conditions is absent ($\Omega = 0$), in the other set it is present ($\Omega = 1$).

What is the nature of the information preprocessing, carried out by the layer of the 32 pyramidal cells, that improved so dramatically the learning abilities of the test dendrites? During the period of the network's learning, pairs of dendrites in each cell taught each other to tune to co-occurring combinations of environmental conditions. Significantly, these combinations of conditions were due to environmental properties a–h. With the dendrites tuned to individual predictable combinations, the entire cells tuned to recognize the presence and absence of those environmental properties. As we discussed above, due to random assignment of input channels on the dendrites, different cells tuned to different properties, but as a group, the 32 cells were likely to discover all eight of them, a through h. In this way, the information about the status of a–h was brought to the surface in the 32 cells' outputs.

With environmental properties a–h represented directly by individual pyramidal cells, the learning task for the test dendrites was simplified: for the α dendrite, for example, the task was changed from that of learning a second-order nonlinear combi-

nation of input channels *1–32* to a much easier one of learning a first-order combination of properties *a–h*. For the Ω dendrite, its task was simplified from a third-order
combination to a second-order one.

If this interpretation of the Figure 2.9B results is correct, then the improvement in
learning by test dendrites should be limited only to *orderly* combinations of environmental conditions. If we were to ask the test dendrites to learn some second- or third-
order combinations of input channels that do not involve properties *a–h*, but are simply random in their composition, then information preprocessing by the 32 cells
should not be of any help, since the cells will not make explicit the "building blocks"
of such accidental combinations. To test this prediction, we trained the test dendrites
on the outputs of the 32 cells, but using different training signals. The α and β dendrites were trained to recognize second-order combinations of elementary environmental conditions (just as objects α and β are such second-order combinations), but
these new combinations had random compositions, not involving any of environmental properties *a–h*. Analogously, the Ω dendrite was trained to recognize a random
third-order combination.

The results are shown in Figure 2.9C, and they confirm our expectation. Note
especially that the learning performance of the dendrite trained on a second-order
combination (shown in the left plot) is much worse than when that dendrite was
trained to recognize a comparable combination (i.e. object α) from the raw information provided by input channels (see left plot in Figure 2.9A). This deterioration of
learning performance is not surprising, suggesting that the 32 cells made some combinations of environmental conditions – the orderly ones – more explicit, in part, by
filtering out accidental combinations.

To conclude this modeling demonstration, we find that the network of cells with
backpropagating dendrites discovers high-order regularities in the environment
remarkably easily, even when the input connections are distributed among dendrites
at random. With their dendrites tuning to co-occurring combinations of environmental conditions, cells learn to recognize the orderly features of the environment that
cause these regular combinations. This in turn enables dendrites in the next network
to discover even more complex, higher-order co-occurring combinations of environmental conditions, as we showed with the α and β test dendrites.

In our modeling demonstration we did not take advantage of any of a number of
readily available, biologically realistic means by which the arrangement of input connections on the dendrites could be optimized. Their detailed demonstration is beyond
the scope of this chapter, but we ought to briefly mention several of them:

- **Lateral anti-Hebbian inhibitory connections**. In real cortical networks, pyramidal cells inhibit each other's somata (they do not do it
  directly, but by activating local inhibitory cells, which in turn inhibit
  other pyramidal cells). If these inhibitory disynaptic connections are
  anti-Hebbian (the inhibitory strength of a connection increases with the
  correlation in behaviors of the presynaptic channel and the postsynaptic soma), then pyramidal cells that happen to develop similar functional properties will also strengthen their inhibitory connections on

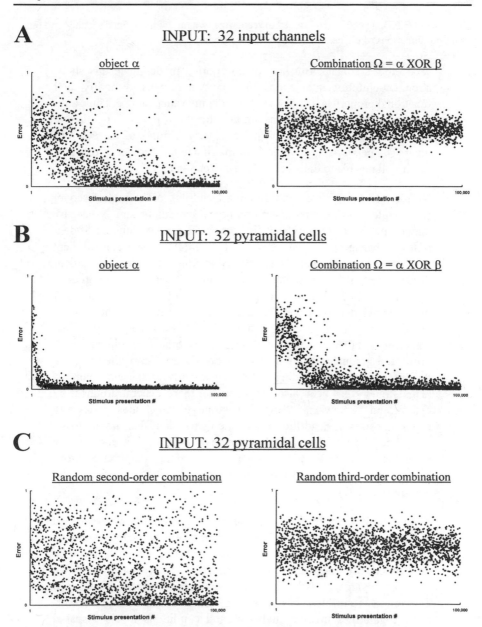

Figure 2.9. Learning performance of two test dendrites (left and right columns) in response to a random sequence of 100,000 input patterns. The error plotted is the magnitude of the difference between the training signal *TR* and the test dendrite's output *D*, showing how well the dendrite predicted the training signal. Plotted are responses only to every 100th input pattern. (A) Two test dendrites, trained to recognize object α (left) and combination Ω of objects α and β, received their inputs directly from the 32 input channels, rather than from the 32 SINBAD cells. (B) The α and Ω test dendrites received their inputs from the 32 SINBAD cells, rather than directly from the input channels. (C) Two test dendrites, with their inputs coming

from the 32 SINBAD cells, were trained to recognize two randomly defined second- and third-order combinations of elementary environmental conditions.

each other's somata. But, as our unreported modeling results show, stronger inhibitory interactions will drive these cells to modify their functional properties to make them less similar than before. Thus, anti-Hebbian inhibitory interactions among somata of pyramidal cells will drive these cells to tune to different orderly features of the environment. The network, as a whole, will maximize the number of regularities it will be able to discover in the environment.

- **Number of dendritic compartments**. Dendrites might fail to discover predictable combinations of environmental conditions simply because they do not have enough dendritic compartments (or, thinking of dendrites as backpropagation networks, they might have an insufficient number of hidden units). Thus, if a pyramidal cell finds that its dendrites cannot, after a reasonably long period of trying, learn to produce matching outputs, then one possible remedy is to add more compartments to its dendrites, i.e., to add more dendritic branches and/or lengthen the existing ones. This might be a basic strategy: pyramidal cells might start with very small and poorly branched principal dendrites and gradually elaborate their dendritic trees until the principal dendrites succeed in finding something orderly in the environment. (This strategy has been also proposed, on more general grounds, by Quartz and Sejnowski, 1997.) This strategy would most efficiently match the sizes of dendritic trees to the demands of their tasks. Possibly related to this idea is the fact that pyramidal cells in the animals reared in behaviorally more enriched (i.e., more complex) environments have more elaborate dendritic trees (reviewed by Quartz and Sejnowski, 1997).

- **Topographic map**. In our last modeling demonstration we connected each cell to all the input channels, but in networks with larger numbers of input channels and, of course, in the cortex this approach would be both practically impossible and functionally detrimental. Instead, we can take advantage of the fact that most orderly relations in natural environments are local in one way or another. For example, lower-order regularities involve environmental conditions in close spatial proximity to each other; consequently, exposing a pyramidal cell in a primary sensory cortex to raw information from distant spatial locations would be useless. In agreement with this observation, afferent connections to cortical areas do not all contact each and every cell but have clear topographic organization (e.g. body maps in somatosensory and motor cortices, retinotopic maps in visual cortex, etc.). These maps are created in middle cortical layers by a host of genetic and epigenetic mechanisms (von der Malsburg and Singer, 1988). From the viewpoint advanced in this chapter, we expect that the mechanisms that control perinatal development and adult maintenance of cortical topographic maps are designed to supply each cortical neighborhood with limited but functionally related information in order to improve its chances of discovering orderly environmental features. Some of these mechanisms, in particular the ones that operate very locally, among neighboring cortical minicolumns, are described in detail in Section 2.4, together with their potential significance for the development of functional properties of SINBAD neurons.

- **Trial-and-error rearrangement of connections on dendrites**. While the initial arrangement of input connections on a cell's dendrites can only be random, it can later be changed, if the dendrites fail to find anything orderly. After a period of unsuccessful learning attempts, few randomly chosen connections might be dropped, while other new connections might be added. This would involve some sprouting of axon collaterals, something that takes place even in the adult cortex (Darian-Smith and Gilbert, 1994; Florence et al., 1998). Dendrites can continue to "experiment" with their input connections until they find co-occurring combinations of environmental conditions.

- **Usefulness of a cell's output**. A pyramidal cell might discover some regularity in the environment, but that regularity might turn out to be inconsequential, of no significance to any other pyramidal cell. (Intuitively, the most useful regularities are those that are most predictive of other regularities, or those that are most relevant to behavior). In that case it would be functionally desirable for the first pyramidal cell to discard the useless regularity and find another one. This mechanism is easy to implement by monitoring the sum of weights of all the connections a given pyramidal cell has on other cells. If no other cell makes use of a given cell's output, then the sum of its synaptic weights will be zero; in that case, after a reasonable period of time the cell can drop and add some of its own input connections, thus forcing its dendrites to find another regularity in the environment that may prove more useful. In the real cortex, monitoring the sum of connection weights might be carried out by well-known trophic signals that presynaptic cells receive from their postsynaptic counterparts (Purves, 1988; Thoenen, 1995); the net trophic signal reaching a cell's soma is indicative of the number and weights of synapses it has on other cells.

- **Multiple networks**. Obviously, a single layer, or network, of pyramidal cells is limited in the complexity of orderly relations it can discover in the environment. However, if its output is fed to another network of pyramidal cells (e.g. primary visual cortical area V1 projecting to V2, etc.), then the higher network will be able to extract higher-order environmental regularities that have as their building blocks the regularities discovered by the lower network. In this way, a series of networks can discover very complex orderly relations in the environment. Higher cortical areas, in turn, provide feedback connections to the lower areas (Van Essen et al., 1992), where they can participate in shaping cells' tuning properties and enable higher-order regularities to help in discovering lower-order ones.

2.4   Cortical Minicolumnar Organization and SINBAD Neurons

So far, we have focused on implementation of the SINBAD design in individual pyramidal cells. In this section we take a step back and look at how such SINBAD cells might be organized in larger functional cortical structures. The functional structures that we consider in this section are cortical minicolumns (Mountcastle, 1978).

Minicolumns are 0.05 mm diameter cords of cells extending radially across all cortical layers; each minicolumn is distinguished from its immediate neighbors by its receptive field and functional properties. A common misconception is that cells located so closely to each other have very similar receptive fields. In fact, most of the experimental literature in somatosensory, visual, auditory, motor, and associative cortical areas is in agreement that while neighboring cortical cells can show a remarkable uniformity in some of their receptive field properties (e.g. stimulus orientation in visual cortex), they can differ prominently in others (for a review, see Favorov and Kelly, 1996). When receptive fields are considered in toto, in all their dimensions, neighboring neurons typically have little in common – a stimulus which is effective in driving one cell will frequently be much less effective in driving its neighbor.

This prominent local receptive field diversity is constrained, however, in the radial cortical dimension. Cells that make up individual radially oriented minicolumns have very similar receptive field properties (Abeles and Goldstein, 1970; Hubel and Wiesel, 1974; Albus, 1975; Merzenich et al., 1981; Favorov and Whitsel, 1988; Favorov and Diamond, 1990; summarized in Favorov and Kelly, 1996). If neighboring neurons have contrasting functional properties, they are likely to belong to different minicolumns.

A number of elements of cortical microarchitecture are responsible for the existence of minicolumnar functional units – among them excitatory spiny stellate cells and inhibitory double bouquet cells (Jones, 1975, 1981; Lund, 1984; Somogyi and Cowey, 1984). Spiny stellates are excitatory intrinsic cells located in layer 4. They are the major recipients of afferent connections from the thalamus or preceding cortical areas; in turn, they distribute afferent input radially via narrow bundles of axon collaterals to other cells in the same minicolumn (see a connectional diagram in Figure 2.10), thus imposing on it a uniform set of functional properties.

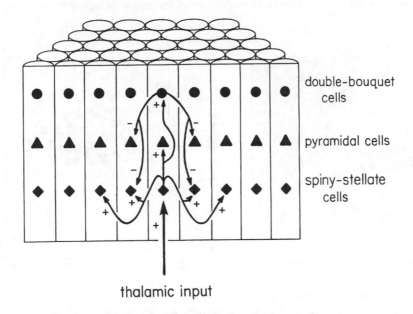

thalamic input

Figure 2.10. Pattern of minicolumnar connections. Shown here is a section across an idealized cortical region, represented as a tightly packed field of cylinder-shaped minicolumns, each containing one representative spiny stellate, pyramidal, and double bouquet cells. Afferent, intra-minicolumnar, and local inter-minicolumnar connections are shown for one minicolumn. Adapted from Favorov and Kelly (1994a). Reproduced by permission from Oxford University Press. © Oxford University Press, 1994.

Double bouquet cells are GABAergic cells with bodies and local dendritic trees in layer 2 and the upper part of layer 3. Their axons descend in tight bundles of collaterals down through layers 3 and 4 and into layer 5, making synapses all along the way on the distal dendrites of pyramidal and spiny stellate cells, but avoiding the main shaft of apical dendrites (DeFelipe et al., 1989; DeFelipe and Farinas, 1992). Due to this arrangement, double bouquet cells are more likely to inhibit cells in adjacent minicolumns rather than in their own (Figure 2.10), and thus they offer a mechanism by which adjacent minicolumns can inhibit each other.

To explain why adjacent minicolumns have noticeably different receptive fields, Favorov and Kelly (1994a, 1994b, 1996) have proposed that during perinatal development each minicolumn is driven by its inhibitory, double bouquet-mediated interactions with adjacent minicolumns (as shown in Figure 2.10) to acquire a set of afferent connections that is different from those of its immediate neighbors. On the other hand, each minicolumn is also driven by the excitatory interactions with a larger circle of its neighbors (see Figure 2.10) to make its set of afferent connections similar to theirs. To satisfy these opposing pressures, minicolumns in local cortical territories arrange their afferent connections in permuted patterns, with shuffled receptive fields. (Shuffling of receptive fields of a local group of minicolumns satisfies the opposing pressures by moving receptive fields of adjacent minicolumns farther apart, while preventing receptive fields of the entire group from diverging too widely).

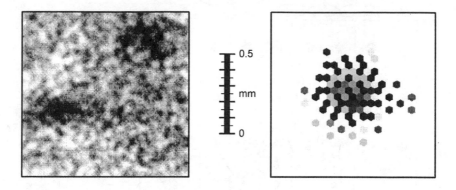

Figure 2.11. Minicolumnar pattern of activation of cat somatosensory cortex: optical imaging of the stimulus-evoked intrinsic signal (left) and modeling prediction (right). The optical image was obtained in the near-infrared range (830 nm; for details about the method, see Tommerdahl et al., 1999). It shows two activated cortical regions, the bottom one in area 3b, the top one in area 3a. Upon closer inspection, the active regions appear as a patchwork of minicolumn-sized spots. Note that these spots tend to be organized in short parallel strings, and that orientation of these strings varies across the activated region. The image on the right was generated by our minicolumnar model of somatosensory network (Favorov and Kelly, 1994a, 1994b). It shows a spatial pattern of active minicolumns, driven by a punctate stimulus. This pattern is similar to the one on the left in that in it active minicolumns are also organized in short parallel strings that run in different directions in different parts of the activated region.

Because of the prominent differences in functional properties among neighboring minicolumns, peripheral stimuli can be expected to evoke spatially complex minico-lumnar patterns of activity in the engaged cortical region, with a mixture of active and inactive minicolumns (Favorov and Kelly, 1994b). This expectation has been experimentally confirmed in studies of stimulus-evoked activity in somatosensory cortex using either 2-deoxyglucose (2-DG) metabolic labeling (McCasland and Woolsey, 1988; Tommerdahl et al., 1993) or near-infrared optical imaging of the intrinsic signal (Figure 2.11). Our modeling studies (Favorov et al., 1994) predict that these minicolumnar activity patterns should be highly stimulus specific and carry detailed information about stimulus features.

To conclude this brief description of cortical topographic organization at the mini-columnar level, it appears that local groups of minicolumns bring together a variety of different, but related sensory information concerning a local region of the stimulus space, with adjacent minicolumns tuned to extract only minimally redundant infor-mation about what takes place in that stimulus space. Because this information comes from a local region of the stimulus space, it is likely to be rich in orderly relations reflecting the orderly features of the outside world. Thus, it appears that local cortical neighborhoods are designed to create local informational environments enriched in regularity, but low in redundancy.

Such local cortical environments are exactly the right informational environments for SINBAD neurons, to be "mined" by them in their search for orderly relations. Each dendrite of a SINBAD pyramidal cell will be extended through a functionally distinct group of minicolumns, exposing it to a unique combination of afferent information. Principal dendrites of the same cell will thus be forced to find different combinations of environmental conditions that are predictably related, as we have shown in the preceding section, and the cell as a whole will tune to the orderly feature of the environment responsible for that relationship.

Thus, we see the progressive elaboration of functional properties of neurons across cortical areas as a repeated two-step process. The first step takes place among spiny stellate cells in the cortical input layer, layer 4. This step involves the development of afferent connections to spiny stellates. This development is controlled by lateral interactions among minicolumns, operating on a number of different spatial scales (inhibitory among adjacent minicolumns, excitatory among farther neighbors, back to inhibitory among even farther neighbors – this is a blend of the classical "Mexican hat" pattern of lateral interactions and the double bouquet pattern we described above; for details, see Favorov and Kelly, 1994b). The task of this development is to organize inputs to the cortical area so that, as we said above, cortical neighborhoods will be supplied with sensory information that is enriched in regularities (by keeping it local but non-redundant).

The second step in the elaboration of neuronal functional properties takes place among pyramidal cells in the upper and deep cortical layers. Spiny stellate cells provide their outputs to pyramidal cells lying above and below them in the same and nearby minicolumns. Pyramidal cells receive from spiny stellates information about some environmental conditions and, in turn, they learn to respond to particular combinations of those conditions, combinations that reflect orderly features of the environment. Pyramidal cells send their outputs to the next cortical area, where the two-step process of building functional properties is repeated, tuning pyramidal cells there to still higher-order combinations of combinations of the environmental conditions, etc.

In conclusion, we view the cortical network as a system that by learning the orderly features of its sensory environment builds an internal model of that environment, along the lines of Craik (1943) and Barlow (1992). This internal model is likely to endow the network with powerful interpretive and predictive capabilities. These capabilities can be used to infer what has happened, is happening, will happen, or what should be done in any particular situation faced by the animal. How such capabilities might be put into practice by a cortical network, and how the entire set of cortical networks composing the cerebral cortex might work together to produce adaptive behaviors are now ready for study.

2.5   Associationism

Mr. Darwin, by a strange inversion of reasoning, seems to think Absolute Ignorance fully qualified to take the place of Absolute Wisdom in all the achievements of creative skill.

Robert MacKenzie, The Darwinian Theory of
the Transmutation of Species Examined, 1868.

The traditional, fundamental, and decisive objection to association is that it is too stupid a relation to form the basis of a mental life.

Jerry Fodor, Modularity of Mind, 1983.

2.5.1   SINBAD as an Associationist Theory

Our proposal is empiricist and associationist. That is, we think the operations of the mind are founded upon a basic neural mechanism that serves to extract regularities from the environment. This extraction of regularity results in the formation of perceptual, and ultimately conceptual structure. By the same token, it allows an organism to make predictions about its environment, and to act accordingly.

Of course this intuitive idea has a long history. The seeds of associationism can be traced back to Plato and Aristotle (Young, 1968), but it first came into its own in 17th and 18th century Britain. The British Empiricists, starting with Locke, opposed the rationalist view that we are born knowing things – we don't have any innate concepts, we don't innately know that anything is the case, and we don't innately know how to do anything. Locke (1700/1978) hypothesized that we acquire knowledge in accordance with principles of association, principles which may be explained by appeal to a small set of "original qualities of human nature" as Hume put it in his Treatise (1739/1978). So, for the British Empiricists, what is innate is not knowledge, but rather a basic mechanism for the acquisition of knowledge from experience.

The first philosopher to provide a systematic theory of the principles of association and their mechanism was David Hartley, in his Observations on Man (1749/1970). He identified two basic principles of association: synchronic and successive. Hartley also incorporated into his system associative links between ideas and movements, and mentioned the influence of reinforcement on associative strength. But neither of these are essential to his associationism (nor was it to Locke's or Hume's). It is unfortunate, then, that the last great research tradition in an associationist vein – Skinner's behaviorism – foundered on its exclusive reliance upon the reinforcement of behavior, and its eschewal of explanations mentioning internal states.

In modern times associationism has become more quantitatively model-oriented, leading to an appreciation of the ability of associative networks to generalize and do similarity-based computations, but it has so far failed to produce an associationist model of cortical learning that is both powerful enough and clearly biologically realistic (Clark, 1993). We believe that our model makes a significant advance in this direction. It is formulated in biologically explicit terms, and thus is eminently experimentally testable. It is self-organizing and, most importantly, it is capable of discovering nonlinear and higher-order regularities. This is the main advance we offer over previous connectionist theories.

2.5.2 Countering Nativist Arguments

Locke's opponents were the rationalists, like Descartes, who posited innate ideas and knowledge. The modern day rationalists are Chomsky, Fodor, and other nativists, who posit genetically endowed concepts and knowledge. In contrast to nativists, our research program is to push the empiricist/associationist line as far as possible before admitting the existence of innate concepts, knowledge, or even biases. This seems a sensible strategy: "It's been there all along" should be a last resort in etiological explanation.

Of course, a radical empiricist position is not plausible for the whole brain; one is born, for instance, knowing how to breathe. Some learning is required even in the autonomic nervous system – (e.g. Hamill and LaGamma, 1992). But this is not likely to be associationist learning. In many subcortical nuclei, for which the nativist case it strong, learning probably involves a domain specific mechanism, similar to those described by Gallistel (1995). This might be called "nativist learning", and it resembles what Chomsky has in mind with his more recent "principles and parameters" account of language acquisition (e.g. Chomsky, 1988).

So in the current proposal, our empiricism amounts to an insistence upon an associationist learning mechanism in a largely equipotential cortex. This is emphatically not equivalent to denying the presence of any constraints on learning. Any associationist mechanism brings in constraints. There is an "argument" against empiricism that goes like this: "You can't get something from nothing. Some mechanism must be innate, which requires that there be some constraints on learning, so empiricism must be false." But no empiricist has ever denied that there are innate mechanisms in the mind. In fact, associationism requires it – recall Hume's appeal to "original qualities of human nature". The empiricist-rationalist opposition concerns not innate structure generally speaking, but rather innate knowledge and concepts. That is, the empiricist denies, and the rationalist accepts, that we have innate knowledge and/or concepts. Empiricist learning theory, i.e. associationism, dictates that knowledge and concepts are acquired through experience.

Together with others (e.g. Clark, 1993; Quartz and Sejnowski, 1997), we believe our proposal for an associationist, empiricist neural model of the cortex provides a plausible alternative to the "hodgepodge of specialized circuits" (Quartz and Sejnowski, 1997) that nativists advocate. However, there are those who feel that associationism is not a genuine alternative, or at least that it is so highly implausible that it is not worth bothering with. They claim that an associationist mechanism plus the environment are insufficient by themselves to account for knowledge we demonstrably have. Therefore, some knowledge must be innate. For example, when children learn a language, they invariably settle on a number of grammatical rules, rules that are independent of any particular language. This universality cannot be explained by an associationist mechanism since the exemplars children have to learn from are compatible with an indefinite number of other rules. Therefore the rules actually used must be innate, and not learned.

Whatever the strength of Chomsky's "poverty of the stimulus" argument in linguistics (see Cowie, 1998 for a trenchant critique), its plausibility is greatly diminished when transferred to the general perceptual domain. First, the argument appeals to the fact that children are not typically reinforced for correct language use, and correction has little effect on their subsequent linguistic behavior – so language acquisition must involve a preset sequence of events that merely unfolds. Though patterns of reinforcement may be relevant to the acquisition of linguistic and other behavior, they are irrelevant to a passive, self-organizing order-extracting mechanism like SIN-BAD. Second, the poverty of the stimulus argument relies on a premise to the effect that children are exposed to a very partial, degraded, and even flawed data set (since adults often use grammatically incorrect speech). The equivalent premise simply does not hold for general perceptual mechanisms. The rules that perceptual AI research proposes as innate are, for example, those used to parse a visual scene to arrive at a three- dimensional model of it. These rules are there to be discovered in an animal's first few glances at a natural scene, in its first tactile interactions with the world, etc. And the world does not make mistakes. (It is interesting to note here a trend in AI research of rejecting sets of rules or algorithms as insufficiently general, leading to models more closely approximating an associationist mechanism [see e.g. Hildreth and Ullman, 1989; Poggio and Hurlbert, 1994].) The evidence does not indicate the presence of innately determined rules that are followed to the exclusion of other rules consistent with the environmental data set. So a poverty of the stimulus argument cannot be given to support complex innate structure in general perceptual mechanisms.

But there is another argument of the same form that is open to the nativist. The argument form, recall, is this: an associationist mechanism plus the environment cannot explain how we have the knowledge we in fact do, so some innate knowledge must be postulated. If the environmental regularities an animal makes use of are "hidden from view", so to speak, because they are higher-order regularities, the animal's knowledge of these regularities cannot be explained by the typical self-organized associationist mechanism. This variation on the standard argument for innateness has been called a version of the poverty of the stimulus argument (Kirsh, 1992). The original version can be conveniently summarized as the underdetermination of a perceptual (or linguistic) system's "theory" of a domain by the evidence available to it. In the variation on the theme, the system's "theory" is not underdetermined. It is simply very difficult to arrive at, since the regularities are higher-order ones. Clearly, SIN-BAD constitutes a promising response to this argument. As we saw in the final modeling demonstration, SINBAD was capable of discovering combinations of combinations of combination of features. In other words, it discovered higher-order regularities, and is capable of finding "type-2 mappings" (Clark and Thornton, 1997). In conjunction with a mechanism for ensuring that all cells choose "useful" regularities (i.e., ones that other cells take advantage of – see "usefulness of a cell's output" at the end of Section 2.3.3), we have an extremely powerful extractor of latent order from the environment.

One further note on the linguistic case: Chomsky's argument relies on the premise that grammar and semantics are independent, in the sense that a child learns her grammar without relying upon semantic information. If Chomsky is wrong about this, and semantic information is actually relevant to learning grammar, then providing a plausible empiricist account of language learning may actually involve surmounting the second rather than the first version of the poverty of the stimulus argument. SINBAD might be up to this task.

Due to the relatively weak order-extracting capabilities of other self-organizing networks (Clark, 1993), many nativists underestimate the power of association. In the quotation at the beginning of this section, Jerry Fodor underestimates associative learning, just like Mackenzie underestimated evolution. Both underestimate the power of a self-organizing system.

We agree with Fodor (1983) that the problem with traditional associationism – Locke, Hume, and Hartley's associations between ideas – is that it is too impoverished, based as it is on correlations of raw sensory inputs. Fodor conceives of a dressed-up, modern, learning-theoretic, computational associationist. This kind of associationist does not limit himself to Hartley's sympathetic vibrations or Hebb's synaptic modifications. He postulates other associative relations: e.g. the logical functions OR, exclusive-OR, and more complex functions are defined, making for a small set of associative relations rather than only one.

Fodor's main complaint about computational associationism is as follows. The associationist is forced to admit more complex operations as fundamental in order to account for more complex mental structure. But an acknowledgment of the complexity of mental structure conflicts with the associationist's account of ontogeny (ibid, pp. 32–34).

In short, as the operative notion of mental structure gets richer, it becomes increasingly difficult to imagine identifying the ontogeny of such structures with the registration of environmental regularities. To put the point in a nutshell, the crucial difference between classical and computational associationism is simply that the latter is utterly lacking in any learning theory.

That is, he thinks the classical associationists had a learning theory (association by the constant conjunction of raw stimuli), but that it could not account for the complexity of mental structure. Modern associationism, by contrast, has a better chance of accounting for the complexity of mental structure, but it is mysterious how a system could acquire this mental structure by extracting regularities from the environment. We think that SINBAD may remove the mystery by showing how nonlinear and higher-order regularities can be extracted from the environment by a simple associationist mechanism.

Neuroscientists are also looking for such a mechanism. In their response to the commentaries on their recent (1997, p. 709) Behavioral and Brain Sciences article, Phillips and Singer ask:

Does unsupervised learning in the cortex discover higher-order variables?

In section 6.4 we asked whether there is any evidence that self-organization in the cortex can discover nonlinear variables such as XOR. No such evidence was offered in the commentaries, nor have we yet found any from other sources. The continued failure of such evidence to appear suggests that reliable discovery of such nonlinear variables may not be a fundamental capability of cortex.

We hope that Phillips and Singer will take heart. If our model is correct, the discovery of nonlinearly separable functions and higher order relations is what the cortex is best at.

2.6   Acknowledgements

This work was supported in part by ONR grant N00014-95-1-0113 and NIH grant NS35222. DR was supported by the Social Sciences and Humanities Research Council of Canada.

References

Abeles, M., Goldstein, M.H., Jr. (1970) Functional architecture in cat primary auditory cortex: Columnar organization and organization according to depth. J. Neurophysiol. 33: 172–187.

Albus, K. (1975) A quantitative study of the projection area of the central and the paracentral visual field in area 17 of the cat: I. The spatial organization of the orientation domain. Exp. Brain Res. 24: 181–202.

Artola, A., Brocher, S. and Singer, W. (1990) Different voltage dependent thresholds for the induction of long-term depression and long-term potentiation in slices of rat visual cortex. Nature 347: 69–72.

Barlow, H.B. (1992) The biological role of neocortex. In: A. Aertsen, V. Braitenberg (Eds.) Information Processing in the Cortex. Berlin: Springer, pp. 53–80.

Brown, T.H., Kairiss, E.W., Keenan, C.L. (1990) Hebbian synapses: Biophysical mechanisms and algorithms. Annu. Rev. Neurosci. 13: 475–511.

Chomsky, N. (1988) Language and Problems of Knowledge. Cambridge, MA: MIT Press.

Clark, A. (1993) Associative Engines: Connectionism, Concepts, and Representational Change. Cambridge, MA: MIT Press.

Clark, A., Thornton, C. (1997) Trading places: Computation, representation, and the limits of uninformed learning. Behav. Brain Sci. 20: 57–90.

Cowie, F. (1998) What's Within. Oxford: Oxford University Press.

Craik, K.J.W. (1943) The Nature of Explanation. London: Cambridge University Press.

Darian-Smith, C., Gilbert, C.D. (1994) Axonal sprouting accompanies functional reorganization in adult cat striate cortex. Nature 368: 737–740.

DeFelipe, J., Farinas, I. (1992) The pyramidal neuron of the cerebral cortex: Morphological and chemical characteristics of the synaptic inputs. Prog. Neurobiol. 39: 563–607.

DeFelipe, J., Hendry, M.C., Jones, E.G. (1989) Synapses of double bouquet cells in monkey cerebral cortex visualized by calbindin immunoreactivity. Brain Res. 503: 49–54.

Deuchars, J., West, D.C., Thomson, A.M. (1994) Relationships between morphology and physiology of pyramid-pyramid single axon connections in rat neocortex in vitro. J. Physiol. (Lond.) 478: 423–435.

Favorov, O.V., Diamond, M.E. (1990) Demonstration of discrete place-defined columns – segregates – in the cat SI. J. Comp. Neurol. 298: 97–112.

Favorov, O.V., Kelly, D.G. (1994a) Minicolumnar organization within somatosensory cortical segregates: I. Development of afferent connections. Cereb. Cortex 4: 408–427.

Favorov, O.V., Kelly, D.G. (1994b) Minicolumnar organization within somatosensory cortical segregates: II. Emergent functional properties. Cereb. Cortex 4: 428–442.

Favorov, O.V., Kelly, D.G. (1996) Local receptive field diversity within cortical neuronal populations. In: O. Franzen, R. Johansson, L. Terenius (Eds.) Somesthesis and the Neurobiology of the Somatosensory Cortex. Basel: Birkhauser, pp. 395–408.

Favorov, O.V., Whitsel, B.L. (1988) Spatial organization of the peripheral input to area 1 cell columns: I. The detection of "segregates". Brain Res. Revs 13: 25–42.

Feldman, M.L. (1984) Morphology of the neocortical pyramidal neuron. In: A. Peters, E.G. Jones (Eds.) Cerebral Cortex, Vol. 1. New York: Plenum Press, pp. 123–200.

Florence, S.L., Taub, H.B., Kaas, J.H. (1998) Large-scale sprouting of cortical connections after peripheral injury in adult macaque monkeys. Science 282: 1117–1121.

Fodor, J. (1983) Modularity of Mind. Cambridge, MA: MIT Press.

Gallistel, C.R. (1995) The replacement of general-purpose theories with adaptive specializations. In: M.S. Gazzaniga (Ed.) The Cognitive Neurosciences. Cambridge, MA: MIT Press, pp. 1255–1267.

Gawne, T.J., Kjaer, T.W., Hertz, J.A., Richmond, B.J. (1996) Adjacent visual cortical complex cells share about 20% of their stimulus-related information. Cereb. Cortex 6: 482–489.

Hamill, R.W., LaGamma, E.F. (1992) Autonomic nervous system development. In: R. Bannister, C.J. Mathias (Eds.) Autonomic Failure: A Textbook of Clinical Disorders of the Autonomic Nervous System. Oxford: Oxford University Press.

Hartley, D. (1749/1970) Observations on man. In: R. Brown (Ed.) Between Hume and Mill: An Anthology of British Philosophy 1749–1843. New York: Random House.

Hebb, D.O. (1949) The Organization of Behavior: A Neuropsychological Theory. New York: John Wiley and Sons.

Hildreth, E.C., Ullman, S. (1989) The computational study of vision. In: M.I. Posner (Ed.) Foundations of Cognitive Science. Cambridge, MA: MIT Press, pp. 581–630.

Hubel, D.H., Wiesel, T.N. (1974) Sequence regularity and geometry of orientation columns in the monkey striate cortex. J. Comp. Neurol. 158: 267–294.

Hume, D. (1740/1978) A Treatise of Human Nature. Selby-Bigge, L.A. (Ed.), Oxford: Oxford University Press.

Jones, E.G. (1975) Varieties and distribution of non-pyramidal cells in the somatic sensory cortex of the squirrel monkey. J. Comp. Neurol. 160: 205–267.

Jones, E.G. (1981) Anatomy of cerebral cortex: Columnar input-output organization. In: F.O. Schmitt (Ed.) The Organization of the Cerebral Cortex. Cambridge, MA: MIT Press, pp. 199–235.

Kirkwood, A., Duden, S.M., Gold, J.T., Aizenman, C., Bear, M.F. (1993) Common forms of synaptic plasticity in hippocampus and neocortex in vitro. Science 260: 1518–1521.

Kirsh, D. (1992) PDP learnability and innate knowledge of language. In: S. Davis (Ed.) Connectionism: Theory and Practice. Oxford: Oxford University Press.

Komatsu, Y. (1994) Age-dependent long-term potentiation of inhibitory synaptic transmission in rat visual cortex. J. Neurosci. 14: 6488–6499.

Komatsu, Y., Iwakiri, M. (1993) Long-term modification of inhibitory synaptic transmission in developing visual cortex. Neuroreport 4: 907–910.

Locke, J. (1700/1978) An Essay Concerning Human Understanding, 4th Ed., P.H. Nidditch (Ed.) Oxford: Oxford University Press.

Lund, J.S. (1984) Spiny stellate neurons. In: A. Peters, E.G. Jones (Eds.) Cerebral Cortex. New York: Plenum Press, pp. 255–308.

Malach, R. (1994) Cortical columns as devices for maximizing neuronal diversity. TINS 17: 101–104.

Markram, H., Lubke, J., Frotscher, M., Roth, A., Sakmann, B. (1997a) Physiology and anatomy of synaptic connections between thick tufted pyramidal neurones in the developing rat neocortex. J. Physiol. (Lond.) 500: 409–440.

Markram, H., Lubke, J., Frotscher, M., Sakmann, B. (1997b) Regulation of synaptic efficacy by coincidence of postsynaptic APs and EPSPs. Science 275: 213–215.

McCasland, J.S., Woolsey, T.A. (1988) High-resolution 2-deoxyglucose mapping of functional cortical columns in mouse barrel cortex. J. Comp. Neurol. 278: 555–569.

McGuire, B., Gilbert, C.D., Wiesel, T.N., Rivlin, P.K. (1991) Targets of horizontal connections in macaque primary visual cortex. J. Comp. Neurol. 305: 370–392.

Mel, B.W. (1994) Information processing in dendritic trees. Neural Comp 6: 1031–1085.

Merzenich, M.M., Sur, M., Nelson, R.J., Kaas, J.H. (1981) Organization of the SI cortex: Multiple cutaneous representations in areas 3b and 1 of the owl monkey. In: C.N. Woolsey (Ed.) Cortical Sensory Organization, Vol. 1. Clifton, N.J.: Humana Press, pp. 47–66.

Mountcastle, V.B. (1978) An organizing principle for cerebral function. In: G.M. Edelman, V.B. Mountcastle (Eds.) The Mindful Brain. Cambridge, MA: MIT Press, pp. 7–50.

Pelletier, M.C, Hablitz, J.J. (1996) Tetraethylammonium-induced synaptic plasticity in rat neocortex. Cereb. Cortex 6: 771–780.

Phillips, W.A., Singer, W. (1997) In search of common foundations for cortical computation. Behav. Brain Sci. 20: 657–722.

Poggio, T., Hurlbert, A. (1994) Observations on cortical mechanisms for object recognition and learning. In: C. Koch, J.L. Davis (Eds.) Large-Scale Neuronal Theories of the Brain. Cambridge, MA: MIT Press, pp. 153–182.

Purves, D. (1988) Body and Brain: A Trophic Theory of Neural Connections. Harvard University Press.

Quartz, S.R., Sejnowski, T.J. (1997) The neural basis of cognitive development: A constructivist manifesto. Behav. Brain Sci. 20: 537–596.

Rumelhart, D.E., Hinton, G.E., Williams, R.J. (1986) Learning internal representations by error propagation. In: D.E. Rumelhart, J.L. McClelland, PDP Research Group (Eds.) Parallel Distributed Processing: Explorations in the Microstructure of Cognition, Vol. 1. Cambridge, MA: MIT Press, pp. 318–362.

Schuz, A. (1992) Randomness and constraints in the cortical neuropil. In: A. Aertsen, V. Braitenberg (Eds.) Information Processing in the Cortex. Berlin: Springer, pp. 3–21.

Segev, I., Fleshman, J.W., Burke, R.E. (1989) Compartmental models of complex neurons. In: C. Koch, I. Segev (Eds.) Methods in Neuronal Modeling. Cambridge, MA: MIT Press, pp. 63–96.

Singer, W. (1995) Development and plasticity of cortical processing architectures. Science 270: 758–764.

Somogyi, P., Cowey, A. (1984) Double bouquet cells. In: A. Peters, E.G. Jones (Eds.) Cerebral Cortex, Vol. 1. New York: Plenum Press, pp. 337–360.

Stuart, G., Spruston, N., Sakmann, B., Hausser, M. (1997) Action potential initiation and backpropagation in neurons of the mammalian CNS. TINS 20: 125–131.

Thoenen, H. (1995) Neurotrophins and neuronal plasticity. Science 270: 593–598.

Thomson, A.M., Deuchars, J. (1994) Temporal and spatial properties of local circuits in neocortex. TINS 17: 119–126.

Tommerdahl, M., Favorov, O.V., Whitsel, B.L., Nakhle, B., Gonchar, Y.A. (1993) Minicolumnar activation patterns in cat and monkey SI cortex. Cereb. Cortex 3: 399–411.

Tommerdahl, M., Delemos, A.K., Whitsel, B.L., Favorov, O.V., Metz, C.B. (1999) Response of anterior parietal cortex to cutaneous flutter versus vibration. J. Neurophysiol. 82: 16–33.

Van Essen, D.C., Anderson, C.H., Felleman, D.J. (1992) Information processing in the primate visual system: an integrated systems perspective. Science 255: 419–423.

von der Malsburg, C., Singer, W. (1988) Principles of cortical network organization. In: P. Rakic, W. Singer (Eds.) Neurobiology of Neocortex. New York: John Wiley and Sons, pp. 69–99.

Widrow, B., Hoff, M.E. (1960) Adaptive switching circuits. 1960 IRE WESCON Convention Record, Part 4 (pp. 96–104) New York: IRE.

Young, R.M. (1968) Association of ideas. In: P.P. Wiener (Ed.) Dictionary of the History of Ideas, Vol. 1. New York: Charles Scribner and Sons, pp. 111–118.

Chapter 3

Performance of Intelligent Systems Governed by Internally Generated Goals

Walter J. Freeman

3.1　Abstract

Intelligent behavior is characterized by flexible and creative pursuit of endogenously defined goals. It has emerged in humans through the stages of evolution that are manifested in the brains and behaviors of other vertebrates. Perception is a key concept by which to link brain dynamics to goal-directed behavior. This archetypal form of intentional behavior is an act of observation into time and space, by which information is sought to guide future action, and by which the perceiver modifies itself through learning from the sensory consequences of its own actions. Chaotic brain dynamics creates the goals, expresses them by means of behavioral actions, and defines the meaning of the requested information. These acts include the making of representations (e.g. numbers, words, graphs, sounds, gestures) for communication to other brains in validation and coordination of experience. The failure of artificial intelligence to achieve its stated aims can be attributed to taking too literally these man-made descriptive representations as the tokens of brain action, whereas in brains there is no information, only dynamic flows and operators.

3.2　Introduction

The dawn of the Information Age with the proliferation of digital computers brought a flurry of optimism regarding the feasibility of using the new devices to emulate the performance of humans in solving complex problems by rational devices. McKenna (Chapter 6, this volume) notes that programs in considerable variety were initiated to explore and exploit this new opportunity for human advancement. As he so clearly describes, the existing devices and programs in artificial intelligence (AI) have failed to match performance qualifications of biological intelligent systems. The questions he raises are, why, and what can be done about it? There are several possible answers. In my opinion the key to finding answers lies in noting the heavy reliance on numbers and number theory in the use both of digital computers and of a broad range of devices that include neural networks, fuzzy logic, probabilistic reasoning, and genetic algorithms as well as the symbolic logic of classical and functional AI. There are no rational numbers in brains, nor is there a foreseeable way of computing with real numbers to simulate continuous time variables (Blum et al., 1989). Dynamical analog devices, which simulate but do not compute, are closer to biological reality, but the development of advanced new analog systems has been largely eclipsed by the seductive "certainties" of digital computation, which are now being developed in the emergent field of "neurocomputation". Neural modelers may object to being grouped with cognitivists and other practitioners of rule-driven symbol manipulation, but neurodynamics as I conceive it is so far from both that they appear to merge, like stars in a distant galaxy that is seen from our own (Freeman, 2000).

Artificial neural networks (ANN) are stable devices that require highly structured, restricted environments for their operation. They are programmed to perform tasks designated by their creators, using information that is provided by their sensors. They are hierarchical in the sense of nested components corresponding to plug-in units in mother boards.

Biological neural networks (BNN) are characteristically unstable, untutored systems that operate in open, unconstrained environments. They devise their own goals and seek information through their sensors that is needed to reach those goals. They

Table 3.1. Some properties that distinguish ANN from BNN.

ANN	BNN
Stationary	Unstable
Input-driven	Self-organizing
Task-oriented	Goal-oriented
Nested circuits	Micro-macro interactive
Computational: - bits - symbols - information	Dynamic: - flows - patterns - meaning
MEMORY	**REMEMBERING**
An object	A process
Representations	Trajectories
Gradient descent	State transition
Retrieval	Construction
Test by matching	Test by action

are hierarchical in a different sense, which requires the co-existence of a myriad of components (atoms, molecules, and neurons) at microscopic levels, leading by their interactions to emergence of macroscopic, self-organizing patterns that constrain the interactive elements (Freeman, 1995). From this perspective the lack of success in emulating BNN with ANN is not due to lack of knowledge about components, such as the various types of neurons, neurotransmitters, membrane receptors, neuromodulators, trigger zones, and brain architectures. The failure is due to inadequate use of known rules of self-organization. Nowhere is this more important than in the endogenous construction and flexible implementation of goals within organisms and, sooner or later, in artificial devices that some researchers hope will offer an improvement over the way that domesticated animals and obedient children perform for humans now.

3.3   Perception as an Active Process

An intelligent device must learn about its body and its environment in order to deal with them effectively. The learning requires an interface between the limited capacity of the organism and the unlimited complexity of the world, and it is done through perception. Classical ANN treat perception as passive. Devices wait for stimuli, and the information impinges on sensors, in which simulated action potentials serve as symbols that are to be manipulated according to logical rules, such as through serial processing into feature detectors, binding into representations of objects, template storage, retrieval, matching by correlation, pattern completion, and task control. BNN are based in active perception. The forms of sensory stimuli are conceived in advance and sought by active looking, listening, sniffing and fingering, with selection for perception only of those limited aspects of the environment which are anticipated and important to the searcher. Every act of perception is based in the goal for which the sensory input is needed to construct meaning. The key problem for neuroscientists and knowledge engineers alike is, how do the goals of intentional activity emerge from the organization of brain systems?

Clearly the work is done in brains by specialized cells. Neurons, like people, are infinitely complex and varied in form and function, but they can only function properly in organized groups. This is shown by their characteristic forms that distinguish them from all other cell types in the body: dendrites and axons. The requirements for high densities of connectivity explain their architectures (Figure 3.1). Extensive branching of the dendrites gives a large surface area for synapses from incoming axons, while providing the necessary channels for synaptic currents to be converged to the trigger zones, from which axon branches diverge to carry the integrated pulse output to many other target neurons. The fibrous tissue infiltrated with capillaries and supporting glia in which the densely packed neurons are embedded is called "neuropil". It forms by the outgrowth of the connecting filaments starting well before birth, and the growth continues throughout life (Figure 3.2). The patterns of connection appear to grow with few specifications by the genome. They are sculpted by learning from experience, with strengthening or proliferation of some synapses and weakening or deletion of others. Cortical neuropil in particular is highly malleable, and it is the main organ of BNN. The distributed sparse connectivity of immense numbers of neurons is the basis for its interactive dynamics. The dense plexus of axons and dendrites particularly in the superficial layer of cortex provides the anatomical basis for its self-organization (Freeman, 2001).

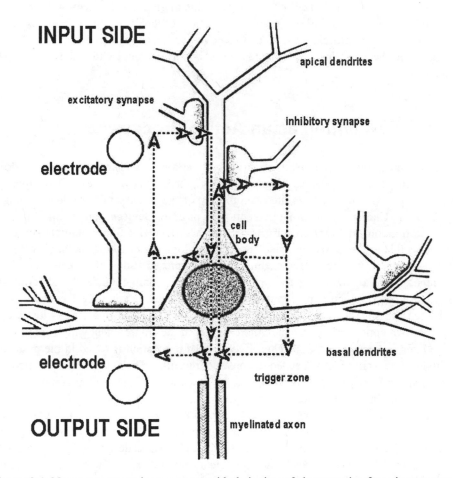

Figure 3.1. Neurons generate loop currents with their sites of electromotive force in synapses and the significant sites of integration at trigger zones. The same currents flow across the extracellular tissue resistance, and the sums of the potential differences from local neighborhoods manifest local mean fields in electroencephalographic (EEG) potentials. Reproduced by permission from Freeman (1992). © World Scientific, 1992.

The growth and maintenance of neurons in neuropil requires that they be continually active. Because they transmit by impulses, their outputs have the form of repetitive pulse trains. Neurons have the capability for relaxation oscillation comparable to

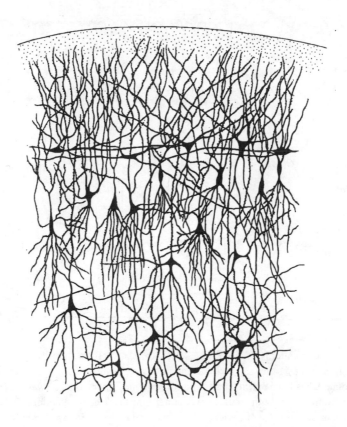

Figure 3.2. The outstanding characteristic of cortical neuropil is the rich connectivity among neurons, which is established early in life and continues to grow throughout maturity. The preferred method of visualizing the connecting filaments is by silver impregnation using the Golgi stain, which selects less than 1% of the neurons. Otherwise, nothing can be seen. Adapted by C. Gralapp from drawings by Ramón y Cajal (1911) with permission from Éditions Maloine.

the beat of the heart, but they seldom show periodic firing. Their pulse trains typically seem random. These firing patterns are due to excitatory interactions. In early development before there are sufficient connections, the neuropil is inactive, manifesting a zero level point attractor (Figure 3.3, "Deep Anesthesia" – this state can be forced onto adult cortex by use of anesthetics). When the growing neuropil reaches a critical level at which the anatomical connection density exceeds unity, a non zero point attractor emerges that homeostatically regulates sustained background activity. The attractor is produced by a first order phase transition that converts a zero point attractor to a repellor (Figure 3.4). The activity of an area of cortex at this stage typically shows nearly "white noise" fluctuation superimposed upon a slowly drifting baseline (Figure 3.3, "Waking Rest").

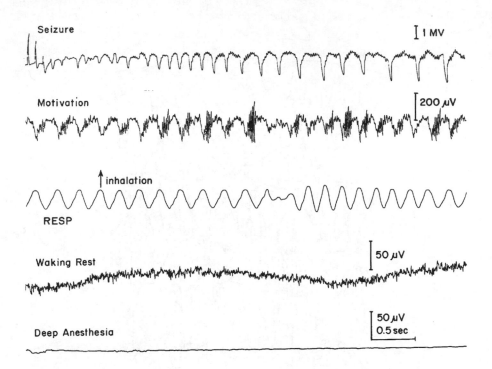

Figure 3.3. Four patterns of EEG time series are shown from the olfactory system. Cortex under deep anesthesia that suppresses interactions replicates the silence of embryonic cortex before connections have reached threshold for sustained interaction. In waking rest the activity resembles white noise. With arousal under conditions that lead to intentional behavior the EEG develops aperiodic oscillations with spectra that tend to the "1/fn-type", where n ~ 2 (Barrie et al., 1996). In the olfactory EEG as shown, the background state is interrupted repeatedly by inhalation under limbic control (middle trace), which manifests a phase transition to a more highly structured form, indicated here by the oscillatory "bursts". Under intense electrical stimulation the olfactory system activity is briefly suppressed, and as it recovers, it enters into a low-dimensional chaotic state that is characterized as a partial complex seizure, a form of epilepsy (Freeman, 1975, 1988). Recovery is by a phase transition directly back to the normal waking state, as shown by the concomitant return of normal EEG and behavior. Reproduced by permission from Freeman (1987). © Springer-Verlag GmbH & Co., 1987.

In early stages of embryological growth the excitatory interactions predominate, and the sustained activity promotes the growth of connections. Around birth or soon thereafter the inhibitory interneurons form connections with the excitatory neurons, providing negative feedback that allows for the induction of a phase transition (Hopf bifurcation) to a limit cycle attractor. Thereafter, local areas of cortex can generate nearly periodic oscillations at the characteristic frequencies determined by their passive membrane time constants and their connection densities (Freeman, 1975). However, during normal activity of the cortex the local areas of neuropil are continually interacting with each other over short and long range connections, so that they seldom sustain periodic activity. As elaborated below, long range connections between areas of cortex sustain oscillators with incommensurate frequencies, leading by further phase transitions to aperiodic activity that is governed by strange attractors.

Learning in the course of normal behavior requires modification of the synaptic connectivity throughout the neuropil. The aspect of learning that I want to concentrate upon is the role played by aperiodic activity in the brain to implement learning. Aperiodic time series are locally unpredictable, but over longer time intervals they are constrained and bounded, so that unlike noise they have internal structure that is not manifested in sharp spectral peaks. It can result when competing components in a dynamic system cannot agree on shared frequencies but cannot escape each other. In systems that are autonomous, noise-free, and low in dimension, that are exemplified by the Lorenz, Rössler and Chua attractors, and that are typically described by a third or fourth order nonlinear differential equation, this type of activity is called "deterministic chaos". The activity in brains as in many other complex systems is not deterministic chaos, because it is driven by and embedded in noise (Freeman, 1996; Freeman et al., 1997). But it is not noise, because it is dependent on cooperation among the vast numbers of neurons which make up the various parts of the brain. This coordination and cooperation shrinks down the phase space of possible activity patterns. Therefore it can be looked upon as stochastic chaos, because the activity arises from and regulates the largely autonomous activity of millions of neurons that are weakly coupled by billions of synapses (Freeman, 1975, 1992, 1996).

3.4   Nonlinear Dynamics of the Olfactory System

The parts of the brain that generate stochastic chaos are not autonomous. They are not only coupled with each other; they are directly involved with the environment, which includes the body as well as the outside world, into which the brain acts by means of its musculoskeletal system, and from which it receives sensory feedback through the several sensory ports. These principles are exemplified by measurements of the electroencephalographic (EEG) activity of the olfactory system (Figure 3.3). This sensory port has three stages of processing following the receptors in the nose: the olfactory bulb, the anterior nucleus, and the olfactory cortex (Freeman, 1975). The bulb receives the receptor input and generates spatial patterns of amplitude modulated (AM) chaotic carrier waves, that is, aperiodic time series having the same irregular wave form over its entire extent. The olfactory cortex receives input from the bulb, performs spatiotemporal integration, and transmits its output into the limbic system, which is the closest approximation in the brain to the CPU of a computer (Freeman, 1995). The anterior nucleus is an intervening control element. Each of these three components is made of neuropil that has millions of neurons, and they interact by both excitatory and inhibitory pathways.

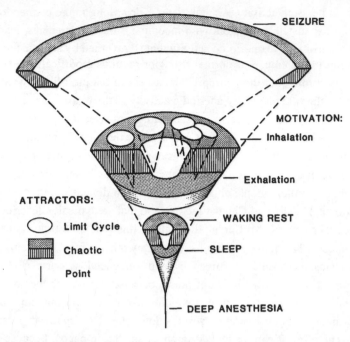

Figure 3.4. A bifurcation diagram illustrates the main states of the olfactory system. Of particular interest is the repeated emergence during inhalation of an attractor landscape with a collection of basins of attraction possibly corresponding to quasiperiodic limit cycles under noise, each basin corresponding to a learned class of stimulus. Reproduced by permission from Freeman (1987). © Springer-Verlag GmbH & Co., 1987.

The EEG during learning in the olfactory system has been studied, first in order to identify the changes in EEG patterns induced by learning, and then to simulate those changes with a model of the olfactory system that consists of a network of ordinary differential equations (Freeman, 1992). The basic module at the nodes of the network is a local population modeled with a linear second order equation with a static sigmoid function. A two-dimensional array of these nodes simulates the distributed interactions of the neuropil; each main part having its characteristic frequency that differs from those of the others. The three oscillators are coupled by long feedback pathways with delays that are dispersive, owing to distributions of conduction velocities and distances within the axonal pathways, which operate as low pass filters. At least one of these long pathways consists of excitatory neurons that excite other excitatory neurons. This constitutes positive feedback, which is responsible for a positive Lyapunov exponent in a model linearized at a typical operating point in state space. Other long pathways contain excitatory neurons that act upon inhibitory neurons, which provide negative feedback that constrains and compresses the activity in the phase space of the olfactory system. So we have continual operation of coupled oscillators which cannot agree on a common frequency, but which cannot escape each other or settle into a point or limit cycle attractor. The resulting activity of the system is governed by a global chaotic attractor (Figure 3.5).

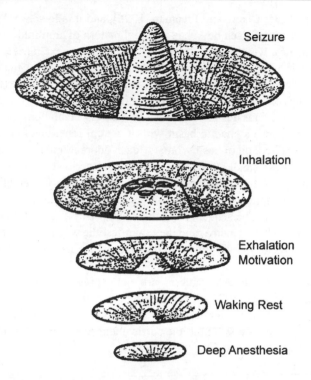

Seizure

Inhalation

Exhalation
Motivation

Waking Rest

Deep Anesthesia

Figure 3.5. The bifurcation diagram from Figure 3.4 is rendered in a perspective drawing, in which the central plateau is seen to be surrounded by a chaotic moat, which is the catch-all basin for "I-don't-know" reactions to novel stimuli. Reproduced by permission from Freeman (1987). © Springer-Verlag GmbH & Co., 1987.

The activity is observed in the EEGs recorded from the three components of the olfactory system, because these extracellular potential differences are established by the flow of current that determines the firing of action potentials, though I emphasize that the extracellular current is epiphenomenal, an indicator and not an operator (Freeman and Baird, 1989). In awake animals that are not engaged in olfactory exploration, the EEGs show that the global activity of the system is governed by a basal attractor that is maintained for minutes to hours, keeping the olfactory system in readiness for action. When the animals are aroused, the EEGs change (Figure 3.3, "Motivation"). The recordings reveal a brief burst of activity, on the order of 50–100 msec in duration, with each inhalation. With continuing respiration the bursts form a sequence of oscillations, which is the manifestation of the control of the olfactory system by the rest of the brain, particularly the limbic system, which is primarily responsible for search strategies using the sensory systems. The repeated bursts with breathing show that search as in sniffing engenders the organization of patterns of neural activity that create the olfactory knowledge in the brain, which is in fact being sought by the brain.

The brief burst of oscillation with each inhalation is concentrated in the gamma range of 20–80 Hertz range, unlike the basal activity that has a broad "1/f" spectrum

(Freeman et al., 1997; Chang and Freeman, 1996), and it is lower in dimension than the background activity. Each burst has a spatial pattern of amplitude and of phase at the center frequency, which differs from each burst to the next. These findings show that each burst is formed by a first order phase transition from the basal chaotic attractor to a more tightly constrained part of the olfactory state space (Figure 3.1). A conjectural image of the state space of the olfactory system (Figure 3.4) shown in perspective (Figure 3.5) is designed to show that during inhalation, a phase transition takes the system from its chaotic basin with a central repellor to open an attractor landscape with a collection of basins, one for each odorant that an animal has learned to discriminate. These small attractors can also be described as wings of the global attractor, in analogy to the Lorenz butterfly with two wings. The amplitude modulation (AM) spatial patterns generated by the wings are then transmitted from the bulb to the olfactory cortex. The sequence of bursts comprises a set of frames of dynamic states, like the frames in a cinema, and they manifest the way in which the limbic system strobes the olfactory environment in order to update its grasp of the ever-changing surround.

I emphasize that these AM patterns are not representations of or information about odorants (Freeman, 1991). They are dynamic operators that enter into other parts of the brain and induce the formation of yet larger cooperative patterns, which in turn solicit interchanges with multiple cortical and subcortical components in each hemisphere.

3.5   Chaotic Oscillations During Learning Novel Stimuli

Most of the bursts occur when the animal is not attending to or searching in the olfactory environment. They nevertheless have value, because they provide the important knowledge that there is nothing of interest at the present time and place, such as the scent of a predator or other sign of unanticipated danger. The burst is a message that is conveyed in the spatial pattern of AM of the chaotic time series in each burst, which has the same wave form in all parts of the olfactory bulb. This is a learned pattern, which is updated whenever the environment changes.

The process of learning involves changes in the synapses which interconnect the excitatory neurons within the bulb and within the cortex, which are co-excited by the sniffing of an odorant on any one trial. On repeated sniffs the cumulative changes in synapses between simultaneously firing neurons leads to the formation of a Hebbian nerve cell assembly for that odorant. The modifiable synapses are not at the site of input from the receptor axons transmitting to the bulbar neurons or at the site of input of the bulbar neurons to the neurons in the olfactory cortex. Those are important sites of modification in respect to signal normalization and logarithmic range compression, but not of associational learning (Freeman, 1975). The modifiable synapses are between the excitatory neurons, which can be described by matrices that are comparable to those developed by Stephen Grossberg, Teuvo Kohonen, Shun-ichi Amari,

John Hopfield, and James Anderson (Shaw and Palm, 1988) in associative networks with sparse, global feedback within each layer of interconnected neurons. The synaptic connections that form the Hebbian nerve cell assembly lead to re-excitation of the interconnected excitatory neurons and an explosive growth of activity by positive feedback. Even though the cell assembly contains only a minute fraction of the bulbar neurons, that explosive growth of activity directs the bulb into a basin of attraction that is defined by learning. Each time a learned odorant is in the inhaled air coming to the receptors, the olfactory system undergoes a state transition, by which the system is driven from its basal attractor and confined to the appropriate basin of the attractor landscape. When the stimulus is withdrawn by exhalation, the system is rapidly released to the basal state by the collapse of the landscape. Each local attractor gives a characteristic spatial AM pattern of the chaotic time series that is manifested in the EEG.

Now what happens when an animal gets a novel stimulus that it has to learn in order to get food or find shelter? The first thing that happens is the expected bursts don't occur (Figure 3.6). Just after each inhalation the limbic system expects a burst to arrive, and if it fails to come, that constitutes a signal to the animal that something is present in the environment, whose nature and significance are unknown. And that something must then be classified. If it is unimportant, it will not be accompanied by food or deprivation, and the bulb will react by diminishing the effectiveness of the output synapses of those neurons that were excited by the unwanted input. This is habituation, which forms an adaptive spatial filter to diminish the impact of irrelevant features in a current environment (Grajski and Freeman, 1989; Gray and Freeman, 1987). The unknown input will be made important when it is accompanied by reinforcement, such as food to a hungry animal, or aversive stimulation of some kind to a threatened animal.

But in either case, during the first episode, there is an absence of an AM pattern of burst activity (Figure 3.6). But there is activity being transmitted that is unstructured. What is happening is that the olfactory system goes to a chaotic basin and sends a message that tells the limbic system, "I don't know what this is, but it may be important!". If it is not, habituation takes place automatically without need for limbic intervention. If it is made important, habituation is blocked, and Hebbian learning begins, which requires coincident pre- and post-synaptic activity. What that deeply chaotic activity provides is the firing which is necessary to drive Hebbian synapses, but without pre-existing structure, which would only result in reinforcement of an existing attractor wing. There must be activity to drive Hebbian synapses, but that activity must be unstructured if a new chaotic basin of attraction is to be formed. So the chaos provides an essential mechanism for making a novel attractor for a novel odor, that is, making something that didn't exist before. That is the essential role that chaos plays. It is not only to bring about an alerting or orienting response, but also to carry out learning by providing the disorder that must precede the emergence of new order.

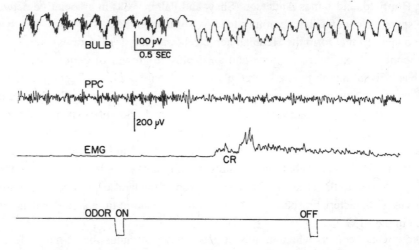

Figure 3.6. EEGs are shown from the olfactory bulb and prepyriform cortex to which the bulb transmits. The EMG (electromyogram) shows the performance of a CR (conditioned response) to a novel odorant after previous training to respond to an odorant CS (conditioned stimulus). The animal gave what is called a pseudoconditioned response to the unknown odorant, and the EEG showed the characteristic suppression of the burst activity, leading us to postulate the deep chaotic moat shown in Figures 3.4 and 3.5. Reproduced by permission from Freeman and Schneider (1982). © Cambridge University Press, 1982.

These synaptic modifications occur during the time duration of a single sniff, which is about a tenth of a second in rabbits and rats. But that only involves a single sample and a small fraction of the neurons that are used for reliable discrimination. Effective learning for generalization requires multiple samples, both with the target odorant reinforced, and without the target but with the background non-reinforced, which allows the animal to learn to discriminate the foreground that is important and to habituate to the background that is not. The acquisition of multiple samples of each odorant is necessary in order to construct a basin of attraction in the bulbar attractor landscape. Each sniff activates only a small subset of the receptors that are tuned to an important odorant, and the selection of the subset varies randomly over sniffs, owing to turbulence in the nose. The incremental formation of a Hebbian nerve cell assembly for that odorant provides for generalization, such that any combination of input from the relevant class of olfactory receptors, even if it never happened before, will put the system into its appropriate basin of attraction and give rise to the AM spatial pattern that conveys the required information to the animal.

3.6   Generalization and Consolidation of New Perceptions with Context

This phase of learning occurs rapidly because the required number of sniffs can take place within a minute or so, as the nerve cell assembly is developed and the AM patterns emerge and are transmitted, reflecting the growth of the new basin of attraction and its chaotic attractor. However, this is only the beginning of the process of incorporation of the novel but now meaningful stimulus into the history of the organism. The consolidation of the learning means not only that a degree of permanence is achieved but, more importantly, that the new learning is knit together with all past experience. The fact that this occurs is revealed in studies of the AM patterns for multiple learned stimuli over periods of days and months. What we find is that whenever a new stimulus is learned and a new attractor forms, the AM patterns for all previously learned stimuli change as well, not by much, but by enough to show that the addition of a new wing to the global attractor jostles all of the others. Therefore, the AM pattern that is the message, which is extracted by the targets of bulbar transmission through a divergent-convergent axonal pathway performing a spatial integration, reflects in some part all of past experience. This is made clear by the progressive, relentless change in AM patterns for all stimuli as daily experience is accumulated. The AM patterns for invariant stimuli are themselves not invariant, because they reflect the changing history, context and significance of the stimuli. What is important for the animals is the meaning of a stimulus, not its particular form, which is constantly changing in any case with each new sample. So we don't have something like pattern retrieval or pattern completion, but instead pattern genesis, which leads now to the conclusion that what the bulb transmits outwardly to the rest of the brain is what it has constructed within itself on the basis of past learning. Another way of saying this is the only thing the nervous system can know about the olfactory environment is what it has created within itself. There is no direct infusion of information from the world.

The state space diagrams (Figures 3.4 and 3.5) summarize the levels of function of the olfactory system. To review, a point attractor is revealed by the suppression of EEG oscillations under deep anesthesia, which is equivalent to "brain dead". This is an open loop state with the gains of the positive and negative feedback set to zero. A point attractor that can be represented as a basin with a deep well, into which the system, if perturbed, will simply drop to inactivity. As the animal recovers to a state of waking rest, the point attractor bifurcates to a central repellor with a chaotic basin around it, in which there is continual flow of activity. With arousal to a motivated state, in which the animal is hungry, frightened, sexually aroused or combative, there is an expansion of this basin, with a change in the attractor landscape to reveal a broad central area within which the state moves in a restless chaotic trajectory.

Input from sniffing destabilizes the olfactory system and reveals the prior existence of a collection of small basins leading to near limit cycle attractors, corresponding to the wings of the global attractor. The input of a known odorant actualizes one of these potential basins, leading to the confinement of the system until the input terminates with exhalation. That is, with each inhalation there is a change in the attractor landscape, so that a collection of basins is made available and the selection of which basin the system will go to is made by the input. Further evidence for this global attractor structure is provided when a very intense electrical stimulus is given to the axonal pathway between the bulb and the cortex. Above a very high threshold, the artificial stimulus drives the olfactory system into an epileptic seizure, which is manifested by an aperiodic spike at about 3/sec followed by a slow wave and a stereo-typic burst. This form of epilepsy is well known in humans as a partial complex or petit mal seizure, and it is accompanied by an interruption of normal behavior known as "absence". The patients are temporarily unconscious, in that they do not respond to social contacts during the episode and afterward cannot remember anything that happened just before or during the seizure.

The EEG manifests a low dimensional attractor, which lasts several seconds and cannot be modified or interrupted by further perturbation, either electrical or behavioral. These properties of stationarity and autonomy allow the epileptiform to be modeled by a low dimensional deterministic chaotic attractor embedded in 3-space, having a fractal correlation dimension between two and three (Freeman, 1988), and a form that is consistent with a collapsed hypertorus (Kaneko, 1986). The epileptic state is also seen during induction of and recovery from the open loop state. The KIII model of the olfactory system indicates that the chaotic state is accessed not by period doubling but by the Ruelle-Takens route – two Hopf bifurcations and the collapse of the torus (Freeman, 1992). Observations on the abruptness of recovery from a seizure show that the state transition from the epileptic state to the normal brain state is by a single bifurcation to a higher dimensional state. The epileptic state is of further interest, in that learning that has taken place just prior to the onset of a seizure is prevented from going on into consolidation. This corresponds to the amnesia observed in human subjects with petit mal epilepsy. It offers further evidence that learning is temporary during the formation or modification of an attractor, and that the heavy work of compiling the fabric of memory is clearly done off-line in some form of batch processing. Many lines of evidence have shown that the hippocampus and related limbic structures in the medial temporal lobes are essential for that work, even though these structures are not the sites of memory storage, as the computer metaphor would hold.

3.7   The Central Role of the Limbic System

Brain scientists have known for over a century that the necessary and sufficient part of the vertebrate brain to sustain goal-directed behavior, a subclass of intentionality, is the ventral forebrain, including those components that comprise the external shell of the phylogenetically oldest part of the forebrain, the paleocortex, and the deeper lying nuclei with which that cortex is connected. These components suffice to support remarkably adept patterns of intentional behavior, after all of the newer parts of the forebrain have been surgically removed from dogs (Goltz, 1892) or chemically inactivated in rats by spreading depression (Bures et al., 1974). Intentional behavior is severely altered or disappears following major damage to the basal forebrain. Phylogenetic evidence for the limbic origin of intentionality comes from observing goal-directed behavior in salamanders, which have the simplest of the existing vertebrate forebrains (Herrick, 1948; Roth, 1987). The three parts are sensory (which is predominantly but not exclusively olfactory), motor, and associational, which contains the primordial hippocampus with its associated septal, amygdaloid and striatal nuclei. The hippocampus is considered to be the locus in higher vertebrates of the functions of spatial orientation (the "cognitive map" of O'Keefe and Nadel, 1978) and temporal integration in learning (the organization of "short-term memory"). These processes are essential, inasmuch as intentional action takes place into the world, and even the simplest action, such as searching for food or evading predators, requires an animal to know where it is with respect to its world, when it got its last signal, where its prey or refuge is, and what its spatial and temporal progress is during a sequence of attack or escape.

The crucial question for neuroscientists is, how are the patterns of neural activity that embody the goals of intentional behavior created in brains? The answer can be provided by extending the models of the dynamics of the olfactory system to the visual, auditory, somatic and entorhinal cortices. The electrical activity of the primary sensory and limbic neocortices of animals that have been trained to identify and respond to conditioned stimuli (Freeman, 1975, 1992, 1995; Barrie et al., 1996; Kay and Freeman, 1998) are closely related to that of the various parts of the olfactory system. As in the case of the olfactory bulb and cortex, neocortical neurons are selectively activated by sensory receptors to generate microscopic activity, and by interactions among the cortical neurons, populations form that "bind" their activity into macroscopic patterns (Hardcastle, 1994; Gray and Singer, 1989; Singer and Gray, 1995). The brain activity patterns reflected in EEGs are macroscopic brain states that are induced by the arrival of stimuli. Each learned stimulus serves to elicit the construction in a primary sensory cortex of a stable pattern that is shaped by the synaptic modifications among cortical neurons from prior learning, and also by the brain stem nuclei that bathe the forebrain in neuromodulatory chemicals (Gray et al., 1986). As a dynamic action pattern it creates and carries the meanings of stimuli for the animal by incorporating the individual history, present context, and expectancy, corresponding to the unity and the wholeness of the intentionality. The patterns created in each cortex are unique to each animal and each experience.

All sensory cortices transmit their signals into the limbic system (Figure 3.7). Much the same kinds of EEG activity as those found in the sensory and motor cortices are found in various parts of the limbic system. This discovery indicates that the limbic system also has the capacity to create its own spatiotemporal patterns of neural activity. I predict that its patterns will be found to depend on past experience and convergent multisensory input but to be self-organized. The limbic system provides a complex array of interconnected modules, that might well serve continually to generate the neural activity that forms goals and directs behavior toward them. Anatomical evidence shows that all primary sensory cortices in mammals direct their transmissions to the entorhinal cortex, and EEG evidence shows that they have the same or similar dynamics. The similarity is essential for the assembly of the inputs deriving from the various specialized sensory ports into the unified Gestalts on which perception is based. Each Gestalt must occur by a dynamic phase transition, in which a complex assembly of neuron populations jumps suddenly from one spatiotemporal pattern to the next. Being intrinsically unstable, the limbic system must continually transit across states that emerge, spread into other parts of the brain, and then dissolve to give rise to new ones. Its output controls the brain stem nuclei that serve to regulate its excitability levels, implying that it regulates its own neurohumoral context, enabling it to respond with equal facility to changes, both in the body and the environment, that call for arousal and adaptation or rest and recreation. I propose that it is the neurodynamics of the limbic system, modulated by other parts of the forebrain such as the frontal lobes and the thalamic and midbrain reticular activating systems, that appears as intentional behavior, and in particular that initiates the novel and creative behavior seen in search by trial and error.

The limbic activity patterns of directed arousal and search are sent into the motor systems of the brain stem and spinal cord. Simultaneously, patterns are transmitted back to the primary sensory cortices, preparing them for the consequences of motor actions (Figure 3.7). This process has been called the "sense of effort" (Helmholtz, 1878), "reafference" (von Holst and Mittelstaedt, 1950; Freeman, 1995), "corollary discharge" (Sperry, 1958), "focused arousal", and "preafference" (Kay et al., 1996). It sensitizes sensory systems to anticipated stimuli prior to their expected times of arrival in the process of attention. Sensory cortical constructs consist of brief staccato messages to the limbic system, which convey what is sought and the result of the search. After multisensory convergence, the spatiotemporal activity pattern in the limbic system is up-dated through temporal integration in the hippocampus. Between sensory messages there are return up-dates from the limbic system to the sensory cortices, whereby each cortex receives input that has been integrated with the output of the others, reflecting the unity of intentionality. Everything that a human or an animal knows comes from this iterative circular process of action, preafference, perception, and up-date (Freeman, 2001).

The limbic system is clearly well situated for controlling the motor systems, because intentional action requires close integration of the central brain state with the conditions in the environment through immediate and close interactions (Taga, 1994; Clark, 1996; Hendriks-Jansen, 1996; Tani, 1996). Its output is directed immediately into two major descending pathways. The lateral forebrain bundle carries activity patterns from the amygdaloid nucleus to the musculoskeletal apparatus to implement overt actions. The medial forebrain bundle carries activity from the septal and accumbens nuclei into the hypothalamus, the head ganglion of the autonomic nervous system and the hypophysis. Moreover, these two descending pathways play major roles in feedback control of the aminergic and peptidergic nuclei in the hypothalamus and brain stem, which are crucial for regulation of affect, learning, and the levels of metabolic energy expenditure, and in oversight through the tract of Vic d'Azyr of the allocation of attention, orienting and awareness resources through the thalamus. In primitive vertebrates the limbic system is dominated by olfaction. Other senses in higher vertebrates have co-opted the basic mechanisms of self-organization. Evolution has also greatly expanded the preprocessing of information in vision, audition and somesthesis, particularly in thalamocortical circuitry, but the algorithms for forming spatiotemporal patterns of chaotic activity appear to have been derived from the olfactory anlage. In the most advanced vertebrates, the limbic system is a complex of interconnected cortical and subcortical structures in the base of the brain forming the medial temporal lobe, which is required for the formation of episodic memories comprising the personal history of each individual, and for control of attention as well as intention (Clark and Squire, 1998).

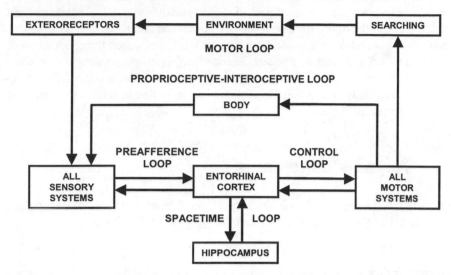

Figure 3.7. In mammals all sensory input is delivered to the entorhinal cortex, which is the main source of input to the hippocampus, and the main target of hippocampal output. Goal-directed action must take place in time and space, and the requisite organ for these matrices is the hippocampus providing "short-term memory" and the "cognitive map". The emergent pattern impacts the brain stem and spinal cord, leading to stereotypic searching movements that are adapted to the immediately surrounding world. Proprioceptive feedback from the muscles and joints to the somatosensory cortex provides confirmation that the intended actions are taking place. The impact of the movements of the body on sensory input is conveyed to the visual, auditory and olfactory systems. All of these perceptual constructs, which are triggered by sensory stimuli and are dependent on prior learning, are transmitted to the limbic system, specifically to the entorhinal cortex, where they are combined. In the genesis of behavior the flow of neural activity is through proprioceptive and exteroceptive loops outside the brain. Within the brain the flow of activity constitutes preafference. When a motor act is initiated by activity descending into the brainstem and spinal cord, the same or a similar activity pattern is sent to all of the sensory systems by the entorhinal cortex, which prepares them for the impact of the movements of the body and, most importantly, sensitizes them by shaping their attractor landscapes to respond quite selectively to stimuli that are appropriate for the goal toward which the action has been directed. These preafferent patterns are the essence of attention as distinct from intention.

3.8   Conclusions

The interface between the finite state intelligent system and the infinite world is active, not passive. The process is self-organizing, by which sequential first order phase transitions lead to formation of complex spatial patterns covering large cortical areas with millions of neurons. The requisite instability and anatomical basis for pattern formation is provided by the dense but sparse connectivity, such that each neuron interacts with its surround, not with small numbers of selected neurons. The trajectory of a perceptual act is not by packets of information that are transmitted from place to place in the brain, starting at sensory cortex and ending at sensory cortex. It is a sequence of phase transitions of cortical neuropil, that solicit the cooperation of varying numbers and sizes of domains in each hemisphere, depending on the complexity of the emergent goal that embeds the process. The process of learning requires chaotic activity as a form of endogenous noise that is subject to control by the limbic system of the brain.

3.9   Acknowledgements

This work was supported by grants from the National Institute of Mental Health (MH 06686) and the Office of Naval Research (ONR-N00014-90-J-4054).

3.10   References

Barrie, M., Freeman, W.J., Lenhart, M. (1996) Modulation by discriminative training of spatial patterns of gamma EEG amplitude and phase in neocortex of rabbits. Journal of Neurophysiology 76: 520–539.

Blum, L., Shub, M., Smale, S. (1989) On a theory of computation and complexity over the real numbers. Bulletin of the American Mathematical Society 21: 1–46.

Bures, J., Buresová, O., Krivának, J. (1974) The Mechanism and Applications of Leão's Spreading Depression of Electroencephalographic Activity. New York: Academic Press.

Chang, H.J., Freeman, W.J. (1996) Parameter optimization in models of the olfactory system. Neural Networks 9: 1–14.

Clark, A. (1996) Being There: Putting Brain, Body, and World Together Again. Cambridge MA: MIT Press.

Clark, R.E., Squire, L.R. (1998) Classical conditioning and brain systems: The role of awareness. Science 280: 77–81.

Freeman, W.J. (1975) Mass Action in the Nervous System. New York: Academic Press.

Freeman, W.J. (1987) Simulation of chaotic EEG patterns with a dynamic model of the olfactory system. Biological Cybernetics 56: 139–150.

Freeman, W.J. (1988) Strange attractors that govern mammalian brain dynamics shown by trajectories of electroencephalographic (EEG) potential. IEEE Trans. Circuits & Systems 35: 781–783.

Freeman, W.J. (1991) The physiology of perception. Scientific American 264: 78–85.

Freeman, W.J. (1992) Tutorial in neurobiology: From single neurons to brain chaos. International Journal of Bifurcation and Chaos 2: 451–482.

Freeman, W.J. (1995) Societies of Brains. Mahwah, NJ: Lawrence Erlbaum Associates.

Freeman, W.J. (1996) Random activity at the microscopic neural level in cortex ("noise") sustains and is regulated by low-dimensional dynamics of macroscopic cortical activity ("chaos"). International Journal of Neural Systems 7: 473–480.

Freeman, W.J. (2000) Neurodynamics: An Exploration in Mesoscopic Brain Dynamics. London: Springer Verlag.

Freeman, W.J. (2001) How Brains Make Up Their Minds. New York: Columbia University Press.

Freeman, W.J., Baird, B. (1989) Effects of applied electric current fields on cortical neural activity. In: E. Schwartz (Ed.) Computational Neuroscience (pp. 274–287). New York: Plenum Press.

Freeman, W.J., Chang, H.J., Burke, B.C., Rose, P.A., Badler, J. (1997) Taming chaos: Stabilization of aperiodic attractors by noise. IEEE Transactions on Circuits and Systems 44: 989–996.

Freeman, W.J., Schneider, W. (1982) Changes in spatial patterns of rabbit olfactory EEG with conditioning to odors. Psychophysiology 19: 44–56.

Goltz, F.L. (1892) Der Hund ohne Grosshirn. Siebente Abhandlung über die Verrichtungen des Grosshirns. Pflügers Archiv 51: 570–614.

Grajski, K.A., Freeman, W.J. (1989) Spatial EEG correlates of non-associative and associative learning in rabbits. Behavioral Neuroscience 103: 790–804.

Gray, C.M., Freeman, W.J. (1987) Induction and maintenance of epileptiform activity in the rabbit olfactory bulb depends on centrifugal input. Experimental Brain Research 68: 210–212.

Gray, C.M., Freeman W.J., Skinner, J.E. (1986) Chemical dependencies of learning in the rabbit olfactory bulb: acquisition of the transient spatial-pattern change depends on norepinephrine. Behavioral Neuroscience 100: 585–596.

Gray, C.M., Singer, W. (1989) Stimulus-specific neuronal oscillations in orientation columns of cat visual cortex. Proceedings of the National Academy of Sciences (USA) 86: 1698–1702.

Hardcastle, V.G. (1994) Psychology's binding problem and possible neurobiological solutions. Journal of Consciousness Studies 1: 66–90.

Helmholtz, H. von (1879/1925) Treatise on Physiological Optics: Vol. 3. The Perceptions of Vision (J.P.C. Southall, Trans.). Rochester NY: Optical Society of America.

Hendriks-Jansen, H. (1996) Catching ourselves in the act: Situated activity, interactive emergence, evolution, and human thought. Cambridge, MA: MIT Press.

Herrick, C.J. (1948) The Brain of the Tiger Salamander. Chicago, IL: University of Chicago Press.

Kaneko, K. (1986) Collapse of Tori and Genesis of Chaos in Dissipative Systems. Singapore: World Scientific.

Kay, L.M., Freeman, W.J. (1998) Bidirectional processing in the olfactory-limbic axis during olfactory behavior. Behavioral Neuroscience 112: 541–553.

Kay, L.M., Lancaster, L., Freeman, W.J. (1996) Reafference and attractors in the olfactory system during odor recognition. International Journal of Neural Systems 7: 489–496.

O'Keefe, J., Nadel, L. (1978) The Hippocampus as a Cognitive Map. Oxford, UK: Clarendon.

Ramón y Cajal, S. (1911) Histologie du Systeme Nerveux de l'Homme et des Vertèbres (Vols. I and II). Paris: Maloine.

Roth, G. (1987) Visual Behavior in Salamanders. Berlin: Springer-Verlag.

Shaw, G.L., Palm, G. (Eds.) (1988) Brain Theory. Reprint Volume. Singapore: World Scientific.

Singer, W., Gray, C.M. (1995) Visual feature integration and the temporal correlation hypothesis. Annual Review of Neuroscience 18: 555–586.

Sperry, R.W. (1958) Physiological plasticity and brain circuit theory. In: H.F. Harlow, C.N. Woolsey (Eds.) Biological and Biochemical Bases of Behavior (pp. 401–424). Madison, WI: University of Wisconsin Press.

Taga, G. (1994) Emergence of bipedal locomotion through entrainment among the neuro-musculo-skeletal system and the environment. Physica. D. 75: 190–208.

Tani, J. (1996) Model-based learning for mobile robot navigation from the dynamical systems perspective. IEEE Transactions on Systems, Man and Cybernetics 26B: 421–436.

von Holst, E., Mittelstadt, H. (1950) Das Reafferenzprinzip (Wechselwirkung zwischen Zentralnervensystem und Peripherie). Naturwissenschaften 37: 464–476.

Chapter 4

A Theory of Thalamocortex

Robert Hecht-Nielsen

4.1 Abstract

This chapter presents the first comprehensive high-level theory of the information processing function of mammalian cortex and thalamus; herein viewed as a unary structure. The theory consists of four major elements: two novel associative memory neuronal network structures (*feature attractor networks* and *antecedent support networks*), a universal information processing operation (*consensus building*), and an overall real-time brain control system (the *brain command loop*). One important *derived* type of thalamocortical neural network is also presented, the *hierarchical abstractor* (which, as with all other networks of thalamocortex, is "constructed" out of antecedent support and feature attractor networks). Some smaller constructs are also introduced. Arguments are presented as to why this theory must be basically correct. **READER WARNING**: The content of this chapter is complicated and almost entirely novel and unfamiliar. Detailed study and multiple readings may be required. Effort expended in learning its content will be richly rewarded. Carrying out computer experiments can be a useful learning adjunct. For simplicity and readability, the theory is presented as fact, without constant recitation of disclaimers such as "it is hypothesized."

4.2 Active Neurons

A key unanswered question about cortical neurons is: under what circumstances will action potentials emitted by a cortical pyramidal neuron influence thalamocortical information processing. It now seems increasingly clear (Fries et al., 2001; Stryker, 2001; Herculano-Houzel et al., 1999; Konig et al., 1995) that these circumstances are rather restricted and that many of the existing hypotheses (such as the pulse frequency coding hypothesis [Cowan et al., 2001; Fain, 1999; Shin, 2001] and the multiple-pulse-arrival synchronization hypothesis [von der Malsburg, 1981; Thorpe et al., 2001]) are probably inadequate. Oddly, one of the earliest guesses regarding this question, Ukhtomsky's *dominanta* hypothesis (Ukhtomsky, 1966; Kryukov et al., 1990; Herculano-Houzel et al., 1999; Borisyuk et al., 1998; Izhikevich, 2001), adumbrated over 70 years ago, seems ever more viable. In Ukhtomsky's core hypothesis, tiny subsets of neurons within cerebral cortex are momentarily mutually phase locked in a brief, episodic oscillation. These neurons form that moment's *dominant*; representing the output of the information processing operation just carried out in that portion of cortex (the dominant presumably automatically emerges as the final step of the processing – in this theory: an *active* token of neurons in the cortical region of a feature attractor network following its operation). A multitude of other nearby neurons are emitting action potentials during this same time period, but they are not involved in expressing the final answer of that moment's information processing (i.e., their signals are essentially ignored). Later, a new dominant is formed as the next information processing step is carried out. These discrete episodic information processing steps can be isolated in time or can be linked into an ongoing chain (as when we concentrate on listening to music). Unrelated neurons which fire rapidly, or even groups of neurons which fire synchronously but which are not phase locked with the applicable dominant over one or more complete oscillation cycles (of which there might be only one to five per typical episode), have no significant effect on information processing.

To avoid plunging into this mystery, and many others related to it, the theory presented here simply assumes that, at each moment at the end of an episode of information processing involving a small *region* (a localized zone of roughly two square millimetres surface area) within any cortical Broca area, a relatively sparse collection of neurons briefly becomes *active*; in the sense that these neurons (a *token*) represent the "answer" produced by the information processing which has been carried out. If no processing is being carried out in that region (the usual situation) or if the processing failed to yield an answer (an occasional outcome), then no token is active within the region – in this case we say that the region is expressing the *null token*. Thus, use of the innocuous word "active" obviates consideration of many details that are unknown at this time (Sejnowski and Destexhe [2000] discusses views of some of the issues involved).

4.3   Neuronal Connections within Thalamocortex

The theory presented here needs to assume the existence of certain specific information signaling channels, or *connections*, between certain groups of neurons. It is clear that almost none of the assumed channels (except perhaps for the random upward connections of the hierarchical abstractor – see Section 4.7) exist in the form of direct axonal connections between the involved neurons. Thus, these channels must be implemented indirectly via signaling through intermediate (*transponder*) neuron populations (as in Hebb's *cell assemblies* [1948] and Abeles' [1991] *synfire chains*) – see also [Chover, 1996; Gerstein, 2001]). Justifying these neuronal communication assumptions would be too large a detour for the purpose at hand, so discussion of this matter is suppressed here. These derivations will be published elsewhere.

4.4   Cortical Regions

The notion of a cortical region is related to that of a *hypercolumn* in the primary visual cortex (Steward, 2000; Nolte, 1999; Mountcastle, 1998). A hypercolumn is roughly defined as that localized volume of cortical neuron somas that are completely responsible for analysing (in terms of the function of the area involved – visual form, color, etc.) one small contiguous "receptive field" of the visual scene being delivered from the retina. Regions have a similar "total responsibility" character but are somewhat larger than hypercolumns. A cortical region is assumed here to have an area of roughly two square millimetres. Regions undoubtedly vary somewhat in size and shape from one cortical area to another. Again, for convenience in describing the first-order picture of this theory, regions will be treated as disjoint and non-interacting; which they are almost certainly not. So, given a total human cortical area of about a quarter square meter, this means that there are roughly 120,000 cortical regions.

4.5   Feature Attractor Associative Memory Neural Network

Each cortical Broca area is assumed here to be massively reciprocally connected with a unique dedicated *area* of thalamus (many other *higher-order* connections between cortex and thalamus also exist [Sherman and Guillery, 2001] – but these are presumably involved in the control of the antecedent support networks discussed below). These reciprocal circuits do not spread much across cortex, and we will envision each region of a cortical Broca area as being reciprocally paired and connected with its own unique thalamic *region* – see Figure 4.1. The existence of these precise massively reciprocal connections has been definitely established for many cortical areas (Sherman and Guillery, 2001; Jones, 1985); but not for all cortical areas, as assumed here. These reciprocal connections are not symmetrical (the number of connections from cortex to thalamus is much larger than the number from thalamus to cortex and the number of glomeruli in a thalamic region is much smaller than the number of neurons in its paired cortical region).

In effect, the cortical surface is exhaustively tiled with regions (each reciprocally connected with its own unique thalamic region). Each of these cortical / thalamic region pairs, along with their reciprocal connections, forms a *feature attractor* associative memory neural network (see Figure 4.1). For simplicity, these feature attractor networks will be treated as disjoint and non-interacting (which they are not).

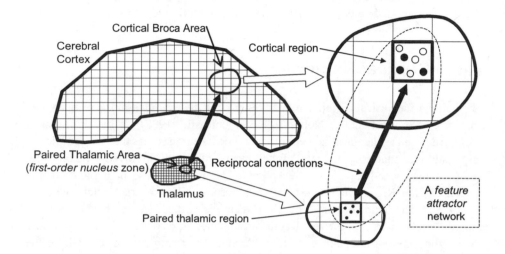

Figure 4.1. Each region belonging to a cortical Broca area (left) is assumed to be reciprocally connected with a dedicated, unique, paired *region* of an area of thalamus (a portion of a first order thalamic nucleus). Each of these cortical / thalamic region pairs and their reciprocal (indirect, in both directions involving cortical pyramidal transponder neurons) connections forms a *feature attractor associative memory neural network*. The main purpose of these networks is to carry out a *feature attraction* function. Given an arbitrary initial collection of differentially *excited* neurons within a cortical region (e.g., the open circle neurons shown in the cortical region) the feature attractor, when *operated*, causes the outputs of these excited neurons to be sent down to the glomeruli of the paired thalamic region. These glomeruli respond to this input and, in turn, send their outputs back to the cortical region – which causes activation of a new, sparse, collection of cortical neurons (illustrated as filled circles in the cortical region of the figure). The rather amazing and key fact is that this new collection of neurons, called a representational *token,* will always be one of a fixed set of tokens (typically a few thousand) established (and frozen) early in life. These tokens are set up during a *critical period* of learning to, in conjunction with other nearby regions, represent objects with minimum overall information loss. This fixed token set is called the *lexicon* of that feature attractor network. The most commonly used lexicon token, the *null token* state of no active neurons within the cortical region, is an indication that either the feature attractor has not been operated recently or that it was operated but that the initial state of the cortical region was not sufficiently similar to any of lexicon tokens. Each feature attractor describes one *attribute* of an *object* (sensory, motor, thought, abstract, conceptual, etc.). The collection of thalamocortical feature attractor lexicons provide a fixed set of terms of reference for describing all of the objects of the world. Learning about the world is possible only because these terms of reference remain largely fixed throughout life. The design of the feature attractor is highly failure tolerant – random death of even a huge fraction of its neurons has little effect on its performance (see Appendix A).

Each feature attractor network is formed during a critical period early in life and then frozen (although they can, under certain circumstances, partially recapitulate their initial formation process – e.g., in response to the appearance of a persistent novel input not well represented by the lexicon). Although the entire cortical surface is assumed to be completely and exhaustively tiled with the cortical regions of feature attractor networks; the corresponding thalamic regions (which are largely mutually disjoint) occupy only a fraction of thalamus. Much of thalamus is involved with control of the cortico-cortical antecedent support networks discussed in the next section. The feature attractor associative memory neural network of a thalamocortical region pair is assumed to be formed during early development (before the antecedent support networks involving the region begin functioning).

Many feature attractor networks have a second mode of operation: relay of "raw data" input entering the thalamocortex from outside (e.g., from sensory or subcortical nucleus sources). When the network is in this mode of operation, these inputs are relayed, largely unchanged, to the cortical region of the network; where they cause "feature detector" neurons to be excited (input from other cortical regions through antecedent support networks can also have this effect). Whenever the feature attractor is operated in its primary feature attraction mode (an operation which typically is completed in about 100 ms) the current excitation state of the cortical region is converted into the nearest matching lexicon token (see Figure 4.1). During this mode of operation the relay of external data ceases.

The manner in which the lexicon of a feature detector develops is not known. Presumably, it involves a competitive and cooperative process across local groups of feature attractors (those which work together to code different attributes of the same objects) which minimizes the overall information loss in the token representation of objects. A simplified feature attractor model which employs uniformly randomly chosen tokens of a fixed size (number of neurons) is discussed below. This is a useful model because the tokens developed by a thalamocortical feature attractor (which are dependent upon the developed feature detector responses of the involved cortical neurons) are likely to have this property (such *quasiorthogonality* [Kainen and Kůrková, 1993] is likely requisite for low information loss).

The dynamics of the feature attractor network (described in more detail below) cause an arbitrary meaningful cortical datum (pattern of feature detector neuron excitement on a region) to be mapped into the nearest matching token of that region's lexicon (the fixed set of non-null *standardized tokens* plus the null token). After a region's critical period (which typically occurs over a span of a few weeks in the first few years of life for a human – although the lexicon for the "tensor analysis" regions may not develop until much later), the lexicon probably only changes under normal circumstances if a radically different input datum (that was never seen during the initial development phase) begins to consistently appear over a long period of time (weeks or months) later in life. When the network attempts to map such a datum into a standardized token the distance is simply too great and the result is the null token and a "reset" of the region results every time. When this occurs over and over again, presumably some partial recapitulation of the critical period is triggered that adds a new standardized token, corresponding to this novel datum, to the lexicon. The stability of the feature attractor network is probably achieved by hard-wiring of the required axonal connections during the network development period. These connections (and the standardized tokens they induce) then normally remain fixed throughout life (as shown below for a simplified feature attractor model, the death of even a huge percentage of the feature attractor's neurons does not harm its function – although the standardized tokens themselves are "eroded".

The key function of the thalamocortical feature attractors is to collectively provide representations for *objects* of all kinds (sensory, motor, thought, abstract, conceptual, etc.) in terms of their large collection of standardized lexicon vocabularies. Each feature attractor lexicon (thus, each cortical region) is responsible for describing one *attribute* of the object. The tokens of the lexicon might be termed *attribute descriptors*. While the representation of any specific object is not usually repeatable (especially at the "lower levels" of representation), the range of its possible expressions is quite restricted (e.g., by transformation group actions). Hierarchies (discussed later in this chapter) are used in many parts of thalamocortex to collect equivalent lower-level representations into more abstract and invariant representations of the same object.

Each episodic operation of the feature attractor network (converting the current cortical response datum to a standardized output token) takes from a few tens to a few hundreds of milliseconds. Even "continuously" ongoing input such as a closely attended soundstream is subjected to this episodic processing (the feature attractors switching back and forth from external input feeding mode to feature attraction mode multiple times per second). As with the episodic saccades of our eyes, we are oblivious to this ubiquitous episodic processing.

In summary, feature attractor networks provide thalamocortex with thousands of fixed "symbolic languages" for describing objects. These stable object descriptors are an essential requirement for accumulation of knowledge over a long period of time. It is interesting to note that several early AI researchers correctly assumed that thinking must be fundamentally symbolic (Nilsson, 1998, 1965; Bender, 1996); although they did not seem to anticipate the vast number of distinct descriptive lexicons involved.

A simplified model of a thalamocortical region feature attractor network is easy to construct and experiment with (it is based on the Willshaw *classical associative memory* [Willshaw et al., 1969; Steinbuch, 1961a, 1961b, 1963, 1965; Steinbuch and Piske, 1963; Steinbuch and Widrow, 1965; Anderson, 1968; Kohonen, 1972, 1984; Anderson, 1972; Nakano, 1972; Amari, 1972; Fukushima, 1975; Anderson et al., 1977; Palm, 1980; Oja, 1980; Amari, 1989]). Thalamocortical feature attractors almost certainly function in a manner similar to reciprocally connected classical associative memories (although their development process has many additional aspects). However, conceptually, my inspiration for the feature attractor came from the Grossberg and Carpenter ART (Carpenter and Grossberg, 1988, 1991; Grossberg and Williamson, 2001; Grossberg, 1999; Grossberg et al., 1997; Grossberg, 1997, 1995, 1987, 1982, 1976), Kosko BAM (Kosko, 1988; Amari, 1974; Haines and Hecht-Nielsen, 1988; Hopfield, 1982; Amari, 1972), Kohonen SOM (Kohonen, 1984,

1995), Anderson BSB (Anderson, et al., 1977), and Fukushima Feedback Cognitron (Fukushima and Miyake, 1978) networks.

As shown in Figure 4.2, there are two *regions* (two-dimensional arrays) of formal neurons, called C and T, representing a pair of connected cortical and thalamic regions making up this simplified feature attractor network model. The C and T regions each have N formal neurons (explained below) and every neuron on each region can send a connection to every neuron on the other.

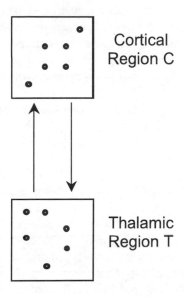

Figure 4.2. A simplified feature attractor associative memory neural network described in the text. The reader may find it useful to build a computer implementation of this network and experiment with it. In particular, this network continues to robustly carry out its feature attractor function even if a large percentage of the neurons involved die. This allows stable object descriptions to be maintained over a lifetime. Such architectures may serve well as a new fundamental basis for artificial intelligence (as well as providing evidence as to the fundamental correctness of this theory).

Before use, this simplified feature attractor associative memory neural network first undergoes the development of its *lexicon*. For simplicity, assume that this lexicon consists of L pairs of standardized tokens (each consisting of one token for C and one token for T, with each token consisting of exactly M active neurons chosen uniformly and independently at random; where $M << N$); denoted respectively by $\{(x_1,y_1), (x_2,y_2), \dots , (x_L,y_L)\}$ (where each token is expressed as a binary N-dimensional vector with exactly M ones). Note that in the biological discussion above there was no mention of a thalamic region lexicon; but this is an important part of the biological construct too. Once this lexicon is constructed (of course, in the thalamocortex this process involves competitive and cooperative interactions between neurons and the lexicon construction process takes place over an extended period of time during early development in response to the local information environment – here we just use a pseudorandom number generator), it is embodied in the network by simply ensuring that the neurons which make up each token in a pair have connections to the neurons making up its mate. No other connections are allowed.

Once this *network formation* process is completed the feature attractor is ready for use for the remainder of life. Of course, the feature attractors in thalamocortex do not have their lexicons chosen first at random and then adapt to them. Their lexicons emerge as a product of a semi-random morphogenic network construction process that this simplified model does not attempt to explain. Note that the other basic thalamocortical network, the *antecedent support network* (considered below) is not like this at all. It continues adjusting during use throughout life in response to changes in the (combined external and internal) *information environment*; and it only stabilizes when the portion of the information environment to which it is exposed is statistically stationary.

The reason random tokens are used (both in this model and probably in the real thing) is that they have many wonderful statistical properties. For example, if we choose a significant number of tokens (e.g., millions) at random, the probability that even two of them will overlap (share common neurons) enough to interfere with each other is extremely small (Kainen and Kůrková, 1993; Ledoux, 2001). The feature attractor can easily separate any standardized token from an activity mixture containing other, lesser activated, tokens and general neuronal background signals. This remains true even after a significant percentage of the neurons of the region randomly die. Of course, as neurons randomly die the tokens themselves become eroded. However, the subset that remains is usually more than adequate to carry on; sometimes even after a century of use. These properties are profound and astounding; particularly in light of the long history of failure of studies of "attractor" neural networks with few if any of these desirable properties. One of the reason for the failure of past studies of this class of networks was a lack of understanding of which properties thalamocortical networks must have. The clear identification of these needed properties is another contribution of this theory.

To operate the simplified feature attractor network we excite a collection of neurons on the C region (it is best to start your experiments with approximately M excited neurons – each excited neuron has an output signal of 1 and all other neurons have an output signal of 0). This *initial datum* or *state* x might, for example, represent the raw output of some previous thalamocortical information processing operation or it might represent the initial cortical response to some external raw data input. The excited neurons of this region C initial state token then transmit their output signal to those neurons of region T to which they are connected (as described above). No other neurons on region C transmit this special signal. The M neurons of region T that receive the largest numbers of these input signals from the neurons of region C are then made *active*. The active neurons of this *T token* then transmit back to region C via the connections established during formation of the network and again the M region C neurons with the largest numbers of inputs are made active. This ends the operating cycle of the feature attractor network. As the analysis of Appendix A shows, this "down and up" process will almost always produce (as its final state on

C) that standard token x_k which most nearly matches the initial C region input in Hamming distance. Unlike many other "attractor" neural networks studied in the past (Amari, 1974; Hopfield, 1982; Cohen and Grossberg, 1983), this network has no "false" or "spurious" attractors, unstable trajectories, or "local minima" (Haines and Hecht-Nielsen, 1988). It rapidly and unerringly converts any initial C region state, as long as it is decidedly closer to one of the lexicon tokens than to any other, into that token. [In mammalian thalamocortical feature attractor networks inputs that are very far from all the standardized tokens are rejected as nonsense and the output is the null token. This simplified network model – see Appendix A – does not exhibit this important behavior.]

If the reader intends to experiment with this simplified feature attractor network, starting with the following parameter values will help ensure success (it is interesting to vary these parameters to explore the properties and limits of this amazingly robust neural network): $N = 10,000$, $M = 30$, and $L = 5,000$. Thalamocortical feature attractor networks have many more neurons than this (at least on the cortical side), but this experiment illustrates many of the basic properties. A key test is to take standardized tokens and alter them (by random additions and deletions of an ever larger number of active neurons) and then quantify (in terms of the Hamming distance of the initial token from each of the standardized tokens – particularly the prototype upon which it is based – as a fraction of M) the network's ability to map each such initial token into the correct standardized token. The results will astound you.

It is also interesting to randomly fail (passive) a growing percentage of the neurons of the C and T regions and then assess performance on the basis of the percentage of the living token neurons which are correctly activated. For a while, no loss of performance ensues (100% of the correct output token's neurons which remain alive are activated); then the structure collapses. Keep in mind that as N and M are increased (to numbers representative of thalamocortex) this robustness increases even further. Experiment with random neuron deaths from 0% to 80% in 5% increments. As the number of living neurons decreases, M must be reduced in proportion.

The use of a rigidly fixed number of active neurons (M) in each lexicon token and a non-local activation rule (activate the M neurons with the largest number of inputs from active neurons) in this simplified feature attractor network are just for computer implementation convenience. More biologically realistic models involving learned graded network threshold control processes (exactly analogous to learned graded muscle control processes) can be built that have roughly this same behavior without any non-local processing.

While real cortical feature attractors undoubtedly have many additional beneficial capabilities; such as an ability to select tokens cooperatively and competitively with adjacent regions; this simplified formal neuron feature attractor network amply illustrates the potential of such a network for describing objects in terms of a fixed lexicon of mid-scale standardized "attribute descriptor words".

4.6   Antecedent Support Associative Memory Neural Network

In the view of Sherman and Guillery (2001), many nuclei of the thalamus are of a *higher-order* nature. Instead of serving to reciprocally connect paired cortical and thalamic regions making up feature attractor networks (which is probably the function of their *first-order* thalamic nuclei), these higher-order nuclei are involved with connections from one part of cortex to another part of cortex. Little is known in detail about these higher-order nuclei. Here it is assumed that they participate in control of the corticocortical antecedent support associative networks described in this section. In effect, it is assumed here that each "region" of such a higher-order thalamic nucleus is responsible for activating a single antecedent support network fascicle connecting two specific regions of cortex. This activation is assumed to take place when "motor" inputs (e.g., arising directly or indirectly from outputs originating in layer 5 of a region of frontal cortex) arrive at the thalamic zone. These "motor" outputs are just like those which emanate from primary motor cortex except that instead of being targeted at motor nuclei of the spinal cord they are instead targeted at higher-order regions of thalamus. Thus, the control of thought is indeed like the control of movement (as discussed further in Section 4.9). As there are roughly a million antecedent support networks (as opposed to about 120,000 feature attractor networks), it is no wonder that a major portion of thalamus is devoted to this function.

Like antecedent support networks, feature attractors can probably also be controlled by "motor" inputs. However, it is also likely that at least some of these networks can also be placed in a *streaming* operating mode where (e.g., sensory) data input to cortex is automatically divided into episodes lasting about 100 ms. At the beginning of each episode, raw input is sent from the thalamic region to layer 4 of its paired cortical region; causing feature detector neurons to respond. Using this response as its initial state, the feature attractor network is then operated (during which raw data input to the cortical region ceases) to map this raw cortical response into an activated standardized cortical token, which can then be transmitted via activated antecedent support fascicles to other regions of cortex and used in thought processes. Then the next episode begins (this process involves some sort of local rhythmic process that is nominally self-sustaining until disturbed). Streaming mode would allow cortical processing of the sensory input data to take place autonomously, with each momentary percept (episodic standardized token) being used by "downstream regions" only when it is needed. Each token is quickly replaced with an updated one. Very likely, such streaming mode operation is often synchronized across many regions at once, and maybe even across hierarchical levels. Streaming operation would explain the feeling that many have had of suddenly "waking up" after many minutes of "automatic driving" down the freeway. During this prolonged period nothing happened to stop the more or less autonomous streaming episodic flow of information through an extensive chain of already-configured linked regions (from the sensorium to the motor areas). In short, it is probably not the case that every activation of every cortical network must be individually commanded. However; setting up such streaming processing chains and carrying out complex processing with significant outcome-dependent branching of network control probably does require individual network control. Just cruising along an uncrowded freeway lane doesn't.

Each thalamocortical *antecedent support* network consists of a pair of cortical regions (the *source* and *target* regions) and a set of unidirectional connections between the neurons of their source and target tokens (implemented using an intermediate population of cortical pyramidal transponder neurons). Each antecedent support network is *operated* using an associated, dedicated higher-order thalamic region which is sparsely connected to the two regions. Only a tiny fraction of all cortical region pairs (far fewer than 1%) are linked by such networks. Two regions which are linked by a antecedent support network are often (but by no means always) also linked by a second such network operating in the opposite direction.

As illustrated in Figure 4.3, the purpose of an antecedent support network is to form unidirectional, weighted, linkages between each token of its source region and certain selected (just for it) tokens of its target region. In other words, the neurons of each source region token (*source token*) are (indirectly) connected to all (well, actually just many) of the neurons of each target region token (*target token*) to which it is to be linked. For each source token the set of target tokens to which it is to be linked are automatically *selected* by using a fixed criterion (discussed below). The strength or *weight* that each of the neuron-to-neuron connections involved in each of these selected token-to-token links implements (they all redundantly implement approximately the same weight — a source of high failure tolerance to random neuron death) is determined by an entirely separate process (also discussed below).

The criterion for selecting a source token - target token pair for linkage (only non-null tokens are linked) is that they appear together (or in very close temporal sequence) at a statistical frequency significantly above chance. The measure of this criterion is the ratio of the frequency of actual observed co-occurrence of the tokens to the frequency with which they otherwise occur together at random. This ratio, which is called the *mutual significance* of the target token in the context of the source token, is denoted by $s(j|i)$, where i is the index of the source token in some enumeration of the source token lexicon and j is the index of the target token in some fixed enumeration of the target region lexicon. Those source / target token pairs with significance above 1 are selected for linkage. The frequencies involved in this criterion are determined over a fixed number of past uses of the network (typically corresponding to weeks or months of real time – although massive rehearsal of a particular token pair during sleep, as in the case of an event memory to be permanently stored, can accomplish the same thing).

Once two tokens are selected for linkage by this significance criterion, a weighted link is formed between each of the neurons of the source token and all (well, almost all) of the neurons of the target token. Once formed, this link is permanent (although its weight is subject to change under certain restricted conditions). Even if the mutual significance of the token pair later drops way below 1, this link remains. This accounts for the remarkable ability of humans and other mammals to recall long-disused information.

Once two tokens are designated for linkage (a function that is almost certainly carried out by the neurons of the *transponder token* which interpose in each communication from the source token neurons to the target token neurons), the common weight of these connections must be determined. This weight, designated w(i,j), is a monotonically increasing sigmoidal function of the quantity p(i|j), the conditional probability of the source token being present given that the target token is present. The frequencies used in calculating this probability are determined over a fixed number of the last times this network has been used. However, once established, the weights of a particular linkage are modified (and then slowly) only when the source token is used. Antecedent support learning conceptually harks back to the continuous on-line statistical adaptation of the classic Widrow (Widrow and Hoff, 1960) ADA-LINE neural network.

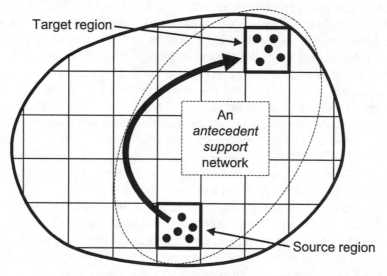

Figure 4.3. An *antecedent support* associative memory neural network. Each (genetically determined) antecedent support network consists of two cortical regions (called the *source* and *target* regions of the network) and unidirectional links connecting the neurons of each source region token (*source token*) with all of the neurons of each of a relatively small number of target region tokens (*target tokens*) especially selected for that source token. The criterion used to select which target tokens to link from a particular source token is that the particular source and target token pair appear together (or in very close temporal sequence) at a statistical frequency significantly above chance. Only tokens which meet this *mutual significance* criterion are linked (again, through an intermediate population of cortical pyramidal transponder neurons). The *weight* of the link from a source token to a selected target token is directly related (via a sigmoidal function) to the logical *support* p(i|j) the appearance of the target token j would lend to the appearance of the *antecedent* source token i (as determined by the statistics of token use over the past so many times the network has been used). Note that this logical implication goes *backwards* from the target token to the source token, even though the link is in the forward direction. All of the (indirect) connections between the individual neurons of the source and target tokens have approximately the same weight (this redundancy confers great neuron death failure tolerance). The antecedent support network is nature's supreme innovation.

Thus, for a given antecedent support network, if the usage frequencies of a particular source token and all of its linked target tokens remain the same then the weights will not change. Also, if a source token becomes largely disused, the weights of its links to its target tokens will not change. It is only in the case where a source token is used and where the frequencies of appearance of its target tokens change that the involved link weights will change. This is why moving to a new house ends up effectively replacing (over a period of many weeks) a large amount of "local space" knowledge from the old place with knowledge relevant to the new place (roll out of bed to the left, reach for the phone on the left); yet childhood memories and most other knowledge is left untouched. Note that the antecedent support network is consistent with Zadeh's classic hypothesis (1965) that human knowledge is "fuzzy."

It is now fairly clear that, under normal circumstances, the axonal wiring of thalamocortex is fixed and frozen for life at an early point in childhood (certainly before the antecedent support networks begin their learning process). As mentioned above, it is probably the transponder token neurons which decide which source and target tokens of a network should be linked – probably using a very simple mechanism involving incremental Hebbian-style (Hebb, 1948) synaptic strength increase on successful use, along with some sort of fixed neuron-wide quota on production of a molecule needed for neurotransmission (such a mechanism would allow above-baseline strengthening of transponder token neuron to target token neuron synapses to occur only if the use statistics of the synapse are above a "chance" level). Once this "decision to link" has been made, then all that needs to happen is that the involved transponder neuron synapses are strengthened. The post-synaptic transduction machinery of the target token neurons which receive neurotransmitter from these synapses are also presumably subject to some fixed neuron-wide production quota. The result is, effectively, a joint pre-and-post-synaptic calculation of the ratio $p(i,j) / p(j) = p(i,j) (1/ p(j)) = p(i|j)$. The first factor is calculated by the strengthening of the pre-synaptic (transponder neuron) component due to Hebbian joint use. The second factor is directly related to the availability of some post-synaptic neurotransduction molecule having (again) a fixed total production rate (which will cause the concentration of this molecule to be inversely proportional to $p(j)$). Thus, although the criterion for linkage would initially seem to require an impossibly large amount of record-keeping by some omniscient entity, this is not the case and the transponder token neurons can probably easily do this job. Also, when transponder neurons do decide a link is necessary, their own involved synapses are automatically strengthened by the correct amounts via a simple Hebbian learning process. Finally, to build the proper link weights, the post-synaptic machinery receiving these strengthened neurotransmitter inputs transduces them with a factor roughly proportional to the inverse frequency of usage of the target token. When the use of antecedent support networks in thinking (described below in Section 4.8) is taken into account, it is easy to conclude that the antecedent support network is the supreme innovation of nature.

For carrying out computer experiments with antecedent support networks, we need a way to approximate the mutual significance $s(j|i)$. If we assume that tokens i and j appear statistically independently of one another when they do not appear together, then the probability of random co-occurrence of tokens i and j is the product of their a priori probabilities p(i) p(j). Using this assumption we get s(j|i) = p(i,j) / [p(i) p(j)]. This is satisfactory for many elementary computer experiments (this formula tends to underestimate s(j|i) for high-frequency source tokens i).

Again, the fact that indirect connections through transponder neurons are sufficient to implement millions of antecedent support networks in human thalamocortex is not established here (these derivations will be published elsewhere). A key side-benefit of this design is that the total number of synapses needed to implement these networks is roughly proportional to the total number of linked token pairs.

The transponder token neuron to target token neuron synapse efficacies (including both the pre- and post- synaptic machinery) used to implement the link weights in thalamocortical antecedent support networks have efficacies which range from 1 (nominal, unstrengthened) to perhaps 30 (maximum antecedent support); corresponding to sigmoided p(i|j) values. Note that these values are well within the capabilities of real synapses (particularly if existing synapses are allowed to "bud" once or twice and create additional nearby augmentation synapses – see Markram, Chapter 5 of this book).

Human cerebral cortex has about 80% x 24 x 10^9 = 19.2 x 10^9 pyramidal neurons, each of which synapses with about 20,000 other pyramid neurons (Mountcastle, 1998; Braitenberg and Schuz, 1998; Nicholls et al., 2001). Thus, there are about 3.8 x 10^{14} pyramidal neuron to pyramidal neuron synapses available for storing knowledge. If approximately 500 of these synapses are used to redundantly store each antecedent support network link weight (this number is based on a separate analysis to be published elsewhere); then each of an assumed million antecedent support networks must have, on average, about 760,000 meaningful token pairs. If an average region has a lexicon of, say, 4,000 standardized tokens then this implies that each source region token has connections to about 192 meaningful target region tokens. This works out to each source region token connecting to about 5% of the target region's tokens. If link weights for all possible token pairs had to be stored, our heads would be required to have roughly 20 times greater volume (2.7 times greater diameter) and mass than they do.

If we assume each token link weight has a dynamic range of five bits, then, factoring out redundancy, the total knowledge capacity of the human thalamocortex is approximately 3.8 x 10^{12} bits or 475 GBytes of information (my desktop computer, which this chapter is being written on, has 400 GBytes of hard disk memory and 2 GBytes of DRAM). A prediction of the theory presented here is that, for most people, a significant fraction of the available brain storage capacity is actually utilized. If we assume that this capacity is filled in the first 32 years of life (a billion seconds), then the average learning information rate is 3,800 bits per second. This is probably roughly correct! During each moment of wakefulness, there are probably hundreds of antecedent support networks active in thalamocortex, many being adjusted. The information appetite of the mammalian mind is truly immense.

A similar, but probably more modest process of rehearsing and imprinting important (i.e., associated with drive and goal states by the limbic system) knowledge and events encountered during the day goes on continually during sleep (Sejnowski and Destexhe, 2000). During each period of wakefulness, every antecedent support network which is used has the token pairs involved *sensitized*; making it much easier to activate these pairs again for several hours. This is the phenomenon of *short-term memory*. As the period of wakefulness wears on, this ever-growing collection of sensitized network token pairs indirectly creates a sort of "background noise" that makes the process of operating antecedent support and feature attractor networks ever more difficult (higher and higher levels of control vigor must be used to get these networks to function). On top of this, the brainstem lowers overall thalamocortical baseline arousal in accordance with the timing of the sleep-wake cycle. Eventually, if wakefulness is deliberately prolonged, the required control vigor levels exceed those which are physiologically possible and sleep becomes non-optional. Sleep (normal, slow wave – not REM) is the (inherently stable) state where the brain command loop (see Section 4.9) is not operating. During the period of sleep, the sensitization of the non-rehearsed token pairs wears off. We then wake up with a blank mental slate; free of yesterday's token sensitizations and their associated "background noise".

Appendix B describes computer experiments employing antecedent support networks that the reader may find useful to implement and carry out.

4.7   Hierarchical Abstractor Associative Memory Neural Network

Although it is not fundamental to the introductory discussion here, it is nonetheless interesting to consider an important type of *derived* thalamocortical neuronal network. By derived it is meant that this type of network is (as are all others in thalamocortex) essentially constructed out of the two fundamental network types: feature attractor and antecedent support networks. The derived network considered in this section is the *hierarchical abstractor* (see Figure 4.4). This network forms, and then uses, unitized *high-level* token representations for one or more groups of lower-level tokens representing more specific objects. In effect, a hierarchical abstractor is simply a collection of antecedent support networks in a particular configuration (with the addition of a special type of connection – which is already present in the involved networks – to be described below).

The focal point of the hierarchical abstractor network construct is a single *higher-level* region. This one region is connected, pairwise, to and from (by antecedent support networks) each of a collection (often tens or hundreds) of *lower-level* regions. Thus, the hierarchical abstractor looks like a star, with the antecedent support networks connecting the central higher-level region to and from the outlying lower-level regions. My development of the hierarchical abstractor was inspired by Steven Grossberg's *instar*, *outstar*, and *avalanche* networks of thirty years ago (Grossberg, 1968, 1982), which had similar functions (*chunking* and *recall*), but entirely different construction and principles of operation.

Besides the antecedent support networks linking the hierarchical abstractor, there are also sparse random upward fascicles of connections from each of the lower-level regions to the higher-level region. These random connections are actually the first random lower-level token neuron to transponder neuron connections that all pyramidal token neurons have. They just happen to be genetically arranged so that all of the lower-level regions to be linked to the higher-level region send some of their connections to it. These connections (which are randomly distributed on the higher-level region) are probably employed by altering the timing of the operation of the higher-level region following operation of the lower-level regions. In particular, the higher-level region is operated with a much shorter delay then when input arriving from transponders is to be processed. In this way, only the direct, sparse, random inputs from the lower-level regions are present at the time that the neurons of the higher-level region are paying attention to their input. When the upward antecedent support networks are to be operated (one from each lower-level region which participates in the hierarchical abstractor network), the higher-level region is operated after a longer delay so that it only responds to the inputs from the transponder neurons excited by the tokens of these networks.

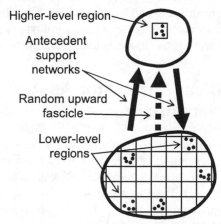

Figure 4.4. *Hierarchical abstractor* networks form higher abstraction level tokens representing groups of lower-level tokens which make up holistic objects (or portions of holistic objects); for example, as identified by using antecedent support networks linking the lower-level regions (this principle is not discussed here). Each lower-level region of the network (there are often tens or hundreds of them) is connected to and from a single higher-level region via dedicated antecedent support networks (one for each direction – shown collectively as solid arrows). The higher-level region tokens arise via sparse random upward connections (shown collectively as a dashed arrow) from each token neuron of the lower-level regions to neurons the upper-level region (these random connections can be selectively and independently activated, when needed). Thus, the higher-level region is connected pairwise with each of the lower-level regions by reciprocal antecedent support networks (which undergo training in accordance with the procedures described in Section 4.6). Hierarchical abstractor networks build and implement an *abstract representation token lexicon* in which each higher-level token typically provides a unitized representation for many, not just one, collections of lower-level tokens. This network allows accumulation and use of knowledge at higher levels of abstraction. By learning exhaustive ranges of possibilities at each hierarchical level, groups of hierarchical abstractors can support *expectation generation*, *action generation*, and *transformation-insensitive object recognition* across a combinatorially explosive set of possibilities (e.g., looking for a **pencil** on a cluttered office table – with all possible colors, poses, lengths, etc. being, in effect, simultaneously admissible).

Obviously, these processes are not started and stopped instantly and must involve region-wide control inputs. Undoubtedly, these control inputs have a close connection with the cognition-related EEG and MEG potentials that have been widely reported and with the fundamental mechanism of neuron activation (see Section 4.2). These random upward connections transmit the signals from the active token neurons on the lower-level regions of the hierarchical abstractor to neurons of the higher-level region. The probability that any active lower-level region token neuron connects to a particular higher-level region neuron can be modeled as a fixed value p.

If it is assumed that K or more of these random input connections from active token neurons of the lower-level regions are sufficient (when the higher-level region is properly prepared for receipt of inputs from the random upward connections) to cause a higher-level region token neuron to become active, then the probability of activation of each higher-level token neuron is a sum of binomial probabilities. If the probability p is chosen correctly, the number of active neurons on the higher-level region will have a roughly Gaussian distribution that is rather narrow. These collections of active neurons are *higher-level tokens* which represent that particular collection of lower-level tokens. Unfortunately, there are many complicated additional mathematical details to this picture, only some of which will be sketched here.

First of all, the higher-level region may or may not already have a token lexicon at the point where it becomes used (forever more) as the "focal point" of an hierarchical abstractor network. In either case, we must presume that some sort of plasticity exists so that, over a period of time, and with a large number of repetitions, the tokens which are formed by the random upward connections (more on how this occurs below) become part of the higher-level region's lexicon. One must also presume that the possibility of, under certain conditions, adding persistent new tokens after the lexicon has been established (as discussed in Section 4.4) exists as well. This is an issue which Grossberg and Carpenter (Carpenter and Grossberg, 1988, 1991; Grossberg, 1995, 1987, 1982) have discussed extensively for their ART network (they term this tension between the need for stable codes and the need to occasionally later revise old, or add new, codes *the stability-plasticity dilemma*). Undoubtedly, as research progresses these issues will be resolved for the feature attractor network of thalamo-cortex (which faces these dilemmas, but probably solves them with as yet undiscovered mechanisms).

A critical issue in the design of the hierarchical abstractor network type is that we want great assurance that the random tokens produced by two different collections of lower-level tokens will not have significant overlap. Here again, the statistics save us – there is no danger of this happening. It is straightforward to show statistically that all pairs of random tokens generated by the above-described mechanism from groups of (assumed) independently generated random lower-level region tokens using random sparse connections will, with very high probability, have low overlap. The surprising thing is that this low overlap happens even when two collections of tokens have a lot in common. For example, what if the lower-level regions (assume there are six of these for a moment) code a sequential sequence of words of text (by equipping each lower-level region with a token lexicon representing the 5,000 most common words in a large English text corpus) and the higher-level region codes the resulting phrase. Then we might have the two (ordered) sets of tokens on these lower-level regions, representing, respectively, the phrases: **it was a beautiful clear day** and **it was a beautiful clear sky**. The astounding fact is that, when p and K and the region sizes (number of neurons) are designed properly (things that genetics can ensure) the higher-level region tokens generated by these two phrases have, with extremely high probability, very low overlap. Thus, the random upward connection mechanism is a way of casting groups of tokens into distinctive standardized tokens of the higher-level region. Traditional thinking would suggest that having such similar objects (two phrases, two very similar visual objects, etc.) represented by essentially "orthogonal" codes is a bad idea. It leaves no leeway or tolerance for insensitivity to small changes in the objects. It also raises the question of how we can deal with the real world, where almost everything we encounter is novel; at least at a raw-data level. Well, as will be seen below, these traditional thoughts are totally wrong. In many ways, this kind of misleading intuition is what prevented this theory from being discovered decades ago.

Another key issue is the problem of variable-size collections of low-level tokens. For example, what about a higher-level region which is supposed to be learning all of the phrases of English (contextually isolated words are considered one-word phrases). As discussed in Appendix C, it is easy to decide when phrases begin and end using antecedent support networks. So, let us say that a hierarchical abstractor network is set up with seven low-level regions (the maximum number of words we will allow in a single phrase). When we sequentially load the phrases of a sentence into the seven lower-level regions of this network (e.g., **Judi [NULL] [NULL] [NULL] [NULL] [NULL] [NULL]** and **said that [NULL] [NULL] [NULL] [NULL] [NULL]** and **she would never again [NULL] [NULL] [NULL]** and **hike [NULL] [NULL] [NULL] [NULL] [NULL] [NULL]** and **over the river and through the woods**); the random upward connections are each time creating a token of a different size. In fact, the binomial statistics are so inflexible that it would not be possible to use a fixed K value and get a reasonable range of token sizes for all of these cases. This indicates the need for a mechanism of K *threshold control*. Here again, we come up against a neuroscience mystery: what mechanism does thalamo-cortex use to accomplish threshold control or its equivalent (note that the simplified trick of choosing the M most activated neurons, as in Section 4.4, would also work – perhaps the same mechanism is used here). The answer is not known. For AI applications, simply varying K in response to the number of active lower-level regions works fine. Finally, as will be noted below, once the higher-level region's lexicon is established it is only rarely necessary to use the random upward connections. At that point (which is presumably usually reached very early in life), many of these issues disappear. So much for the random upward connections.

Every time the random upward connections are utilized and a lexicon token is excited, the feature attractor network of the higher-level region is operated. At that point, a token is active on the higher-level region which represents the collection of tokens being expressed on the lower-level regions. Typically, both the upward and downward antecedent support networks of the hierarchical abstractor are then operated and undergo training at this point. The upward connections learn to map from the group of lower-level tokens to the higher-level token and vice versa for the downward connections.

One of the ways hierarchical abstractors can be trained is by a method (implemented by a particular *thought process* – see Section 4.9) called *invariant discounting* (a method which is very common in visual perception). For example, imagine that a human baby is holding an object and changing its pose while staring at it (babies learning to see have eyes which can only clearly see objects roughly within their radius of grasp). Now consider a higher-level region in a secondary or tertiary visual region being fed (within a hierarchical abstractor network) by activated tokens from the next "lower" visual level. At some random point in this process the baby activates a thought process that activates the random upward connections from lower-level tokens describing the local visual appearances, if any, of the object at that moment, to this higher-level region on the next visual level. This higher-level excited token is then converted to a standardized token by operating (training) the feature attractor of that region. The antecedent support networks linking this standardized higher-level token to the many lower-level tokens, as well as those going upward, are then activated (which puts them through an increment of antecedent support learning). The thought process then activates each of the lower-level regions – to update their visual input (i.e., they receive another blast of visual input, either directly from the LGN or from their own lower-level regions, and their feature attractors then operate).

During each of many moments that immediately follow this initial event, the thought process repeatedly carries out these steps: the lower-level regions are updated and the upward and downward antecedent support networks are activated (which causes learning). However, the token on the higher-level region is <u>not</u> changed during this process. So, it comes to represent not just one of these ensembles of lower-level tokens, but many. This higher-level token represents some portion of a fixed visual object seen across a range of poses. After a short time (or if the object is no longer being viewed); this thought process is stopped. The whole thought process is repeated many thousands of times over a period of a few months.

The net result of this learning process is the development of a multitude of hierarchical abstractors which can progressively respond to visual objects in a more and more pose-insensitive way. My colleague Cindy Elder and I carried out a primitive experiment with a simplified version of such a visual system [unpublished work]. The system did indeed develop a hierarchy of visual token lexicons which were progressively more pose-insensitive. This experiment employed a simple visual environment in which capital Latin alphabetic characters would randomly move about on a plane, smoothly expanding and contracting in scale and rotating left and right by $\pm 15°$. An "eyeball" would periodically fixate on the approximate center of the character and use this visual input to feed two-dimensional Gabor logon feature detector neurons in primary visual regions (these would then feed the hierarchical abstractors).

In the action realm, a higher-level token intended as the code for a temporal sequence of individual lower-level actions is left on during successive rehearsals of an *output token sequence* (employing an *action elaborator network*; another type of

derived network, related to the hierarchical abstractor, not discussed here). Antecedent support learning then causes a sort of *average performance* to be learned. Slow or
dry (candidate actions generated but not executed) rehearsals can be used to further
strengthen the average performance and fade out learning which occurred during bad
performances.

Just as the vision and action examples above are quite different, so sound processing is different again (for an example see Sagi et al., 2001).

Hierarchical abstractors have a multitude of possible uses. Investigation of these
structures has only just begun.

4.8   Consensus Building

This section offers the surprising conjecture that, in essence, besides feature attraction, there is only one information processing operation used in thalamocortex: *consensus building*.

Each individual consensus building process involves three elements: a selected
set of *assumed facts*, each expressed as a single active token on a *constraint region*;
an *answer region* upon which a single (possibly null) token is to appear as the *output
answer* of the consensus building process; and a set of antecedent support networks
(each viewed as a separate *knowledge base*) each linking a constraint region to the
answer region. The consensus building process involves a novel answer region oper-
ational mode called *honing*, which is described below. *Thought processes* are stored
combinations of one or more consensus building operations (often with overlapping
elements – see Figure 4.5).

To carry out consensus building, the knowledge bases (antecedent support networks linking one of the assumed fact tokens on its constraint region to the answer
region) are operated, one at a time, in a rapid sequence. By this it is meant that the
neurons of the answer region receive inputs from each assumed fact token, through
its knowledge base, in isolation, but in rapid succession. The order in which these
inputs occur does not matter (the order is coded into the thought process). The first
such input to the answer region excites each target token neuron that is linked from
the assumed fact token i_1 which is active on the involved constraint region. The *level*
of this excitation is the fixed (sigmoidal) monotonic (linkage weight) function of
$p(i_1|j)$, where j is the token to which the target region neuron belongs (for the
moment, assume that each neuron belongs to only one token). This is the first stage
of the honing process. At the next stage, inputs to the answer region neurons arrive
from the neurons of the second assumed fact i_2 (on the second constraint region) via
its knowledge base. The input excitation to the same neuron considered above at this
second stage is clearly the same fixed (sigmoidal) monotonic function of $p(i_2|j)$. At
the instant this second input excitation is delivered to the answer region, the neurons
of the region undergo *honing*. Honing sets the new excitation level of each neuron to
the lesser of its current excitation level and the new input excitation level. This process of input from the next answer region in the sequence and honing (just as in the

second stage) is then repeated, one stage at a time, until all of the remaining answer regions have provided input. The final excitation levels of the answer region neurons are then used as the initial input for that region's feature attractor; which is operated. This concludes the consensus building process. The token active on the answer region at the conclusion of its feature attraction operation is the *output answer* of the consensus building process. In effect, the output answer of the consensus building process satisfies:

> **The Law of Mammalian Thought**: For each consensus building process of a thought process, the answer region being honed ends up expressing that non-null token j (or the null token if no such consensus token j exists) which maximizes the quantity MIN[$p(i_1|j)$, $p(i_2|j)$, ... , $p(i_k|j)$], where i_1, i_2, ... , i_k are the assumed fact tokens used with the knowledge bases impinging upon this region.

In words, this law says that (if there is a non-null answer) **the answer token is that logically viable attribute descriptor which most strongly supports the weakest item (assumed fact) of evidence in its favor**. By *logically viable* it is meant that this answer has established knowledge base links from all of the assumed facts involved in its activation (which implies that it is above the minimum level of mutual significance with respect to each and every one of these facts). Thus, any (non-null) answer is <u>guaranteed</u> to be a logical consequence of <u>all</u> the assumed facts in the context of <u>all</u> the applied knowledge. And, among all such logical consequences (if any there be), it is the one which most strongly supports the weakest item of evidence in its favor (i.e., it has the maximum minimum $p(i_k|j)$). As cognitive scientists and philosophers become aware of this universal characterization of mammalian thought it will undoubtedly suggest many concrete tests and yield many new insights into the nature of the mammalian mind.

A thought process (see Figure 4.5) can be viewed much like a fixed algorithm with variable inputs: plug assumed facts into the constraint regions and then carry out the thought process: the output will be the consensus answer for that specific collection of facts. Normally, as we experience life, few of the assemblages of assumed facts used in thought processes have ever appeared before; i.e., almost all situations we encounter in life are *novel*. Consensus building is the universal fundamental mechanism of thalamocortical information processing. It is a complicated departure from the traditional simple AND, OR and NOT deductive logic operations which form the basis of today's computer information processing and AI. Consensus building is us.

The consensus building process implements what might be called an *ALL* (a sort of 'antecedent AND') cognitive operation. The other critically important basic cognitive operation is the *ANY* (an 'antecedent OR'); where some maximal *viable subset* (but <u>not</u> the full set) of the assumed facts of a consensus building thought process, considered in isolation without the other original assumed facts, is sufficient to reach a non-null consensus output answer. The token which is chosen as the ANY answer is that with maximum minimum $p(i_k|j)$ among the subset of assumed fact tokens employed. Thus, once the viable subset is identified, ANY is just normal consensus building using this subset of assumed facts. The *piling and reflection* thought process by which this maximal viable subset is determined is not discussed here.

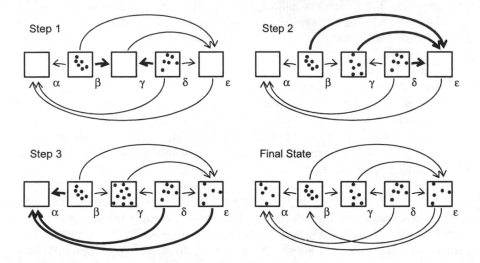

Figure 4.5. Hypothetical thought process example. This example consists of three consensus building processes. At the beginning, constraint regions beta and delta have assumed fact tokens being expressed on them. The other three regions are blank. The first consensus building operation takes the inputs from constraint regions beta and delta and sends them through the knowledge bases (antecedent support networks) linking those regions to region gamma, the answer region of this first consensus building process; which is *honed* during these inputs (see text for details of answer region honing). Notice that this consensus building process would be the same no matter what assumed fact tokens were expressed on regions beta and delta. Thus, thought processes, just like many computer algorithms, are data-independent. This is critical, because most situations we encounter in life are novel; meaning that the assumed facts being used in the thought process have never been seen together before. The second consensus building step of this example thought process takes the tokens on regions beta, gamma, and delta and hones region epsilon to yield an answer token. Finally, in the third step the last four regions are used in a consensus building operation to yield an answer token for region alpha. A key fact is that tokens on certain cortical regions actually code the operation of specific feature attractor and antecedent support networks (not to mention movements). Thus, the answers produced by thought processes can themselves (conditionally) trigger subsequent thought processes (and movements). Thus, thinking is capable of universal computation. A three-consensus-building-step thought process like this one can be carried out in a fraction of a second.

If an ANY operation yields an answer token, this is at least confirmation that the candidate maximal viable subset is a set of facts that are all logically consistent with this conclusion. To witness an ANY operation in your thalamocortex, just answer the question of which sense the word <u>train</u> is being used in the sentence: The dress had a seven foot train. Presumably, an ANY is used to activate a *word family* (which indicates word usage sense) token on an answer region as a consensus based on inputs from the maximal viable subset <u>dress</u> and <u>foot</u>. Although the ANY works much of the time, it can fail to yield an answer and it can yield inconsistent answers (e.g., after blanking an answer the ANY can yield a second different answer with a totally different maximal viable subset). Applying an ANY operation multiple times to essentially the same input (which has multiple interpretations, each of near-equal validity) can cause the answer token neurons to become fatigued and, after a while, result in a different answer token, sometimes with a totally different maximal support set, being supplied as the answer. Over further applications of the thought process these neurons fatigue and the answer switches once again. This phenomenon is commonly referred to as *gestalt switching* (e.g., as with the Necker cube or other ambiguous images).

The neuron *blanking* operation can be used to implement a general-purpose NOT operation in conjunction with a consensus building process (it can be used in other ways as well). First, each assumed fact token which is to be NOTed has its output sent through its knowledge base to the answer region. Neurons corresponding to all tokens which logically support these assumed fact tokens are excited by this input. All of these answer region neurons are then blanked and the consensus building process immediately proceeds. The result is a token (if any there be) which logically supports the affirmative assumed facts but which does not logically support the NOTed fact tokens. To demonstrate NOT, think of a car that is not a Toyota.

Computer experiments with consensus building are described in Appendix C.

4.9   Brain Command Loop

The network operation sequence which encapsulates a thought process (see Figure 4.5) in a data-independent way is stored, selected as a candidate for possible execution, evaluated for appropriateness, and executed by the *brain command loop* in exactly the same way, using exactly the same mechanisms, as the muscle contraction sequences of a movement. The basic idea is that many cortical regions have their active feature attractor output tokens sent to subcortical destinations. These outputs directly or indirectly cause cortical (feature attractor or antecedent support) network operations and/or muscle contractions to take place. The only difference between a muscle contraction and a network operation is the destination (motor nuclei of the spinal cord or subcortical brain nuclei, respectively). Each output has the same character (a time-varying level of output vigor). Thus, in this view, many non-motor areas of cortex have a 'motor' function (Sherman and Guillery, 2001).

At each moment of time, a significant fraction of cortical regions (e.g., many in frontal cortex) are being supplied with inputs describing the current *world state* (both internal and external). When the world state dictates, these antecedent support network's target regions (these networks themselves being controlled by thought processes) generate (via consensus building operations) candidate *actions* (a general unitized term for movements, thoughts, and combinations of the two); many types of which are immediately *vetted* for consistency with current drive and goal states (e.g., by being sent to the basal ganglia) and, if acceptable, flow right through to be *exe-

cuted. This *brain command loop* runs continuously throughout wakefulness, in many parallel streams, involving many subcortical structures.

Many thought processes act as conditional execution tests (sort of like a CASE statement in a computer program). The next action to be launched is decided on the basis of the current state of the world (e.g., on the outcome of the last step of an ongoing thought process). Using action elaboration hierarchies (see Section 4.7), complex thought processes are represented at higher levels by single tokens. Invoking such a thought process only requires the expression of that one token on its region (a result that may first require a command to operate the feature attractor of that region to be vetted before execution). The lower level regions then proceed to elaborate it and carry it out.

Besides elaboration, many action processes must also be *instantiated.* This means that the thought processes involved in creating the final action process are subject to modification by the moment-to-moment state of the world. Instantiation is carried out be incorporating real-time world-state representation tokens as assumed facts in detailed action generation consensus building processes. A subtle, but profound, aspect of this is that, by virtue of the region-pair character of cortical knowledge, almost all instantiations involve combinations of influences that have never before been seen together – and yet the result is automatically appropriate. For example, if I am throwing a baseball while leaning sideways and rotating (which I have never before done together), the compensations for leaning sideways and for rotating are simply "added" in accordance with the mammalian law of thought. Unlike a situation where we are trying to combine logical knowledge about some sensory object; where the law obviously will work well, in this situation this rule of knowledge combination would seem inappropriate. However, it isn't! Why? Because the representational system in which cortical motor commands are coded (postural goals described in a very special coordinate system) is specifically designed to allow myriad different corrections to be "added" by thought processes. This scheme only works over a certain range of kinematic and dynamic situations (try throwing a baseball sometime while hanging upside down – you can't). However, for the majority of everyday motor control applications it works very well (particularly when thalamocortical motor outputs are combined with real-time "critic" feedback corrections arriving at motor cortex from the cerebellum). Phylogenetically, the motor control system and the thought control system probably arose together. They are two aspects of a single *action control system.*

The involvement of subcortical nuclei in the brain control loop is massive. This involvement includes not only vetting of candidate actions but also implementation of those actions. The brain control loop runs continuously throughout the period of wakefulness in many parallel streams; some independent from one another, others launched together by the same thought process. The time it takes for a candidate action to be sent to the basal ganglia, checked for consistency with current limbic and hypothalamic drive and goal states, and sent on through (e.g., to the brainstem or thalamus) for *execution* is often less than 100 ms. Some basic actions are not even subject to vetting and some ongoing thought processes are allowed to keep looping for an extended time with little supervision. Note that, unlike the "cerebral" processes of thought, the release of candidate actions for execution is controlled by subcortical structures seemingly little changed since the age of reptiles. This accounts for the classical psychological view that, no matter how noble our conscious intentions, our spontaneous behavior is selected by our "reptilian mind" in direct response to the drive and goal states that are really active (as opposed to the states that we intellectually wish were active). Figure 4.7 illustrates the brain command loop.

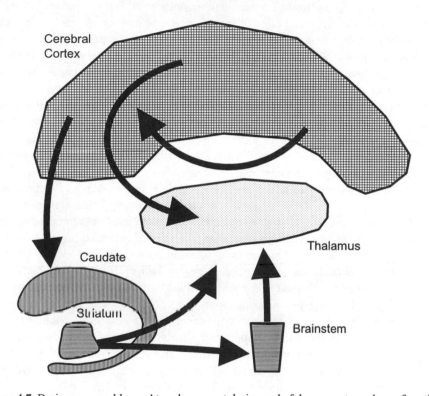

Figure 4.7. Brain command loop. At each moment during wakefulness, vast numbers of cortical antecedent networks carry out thought processes based upon newly received inputs from both cortical and extra-cortical sources. The non-null answers produced by these candidate action generation processes are sent to subcortical structures (e.g., those shown in the figure) for vetting and/or execution. Many of the inputs to thalamus are inputs to be passed along to cortical regions via the external input relay mode of operation of feature attractor networks (these are not shown here). However, many other inputs to thalamus (e.g., the ones shown here) serve to operate feature attractor and antecedent support networks. Along with the signals to spinal motor nuclei, these network operation signals are the main products of the many streams of the brain command loop. Although the brain command loop is comprised of many separate streams that usually function in a largely asynchronous manner; music and dance can, under certain conditions, entrain almost all of these streams and phase lock them to a single beat. It is this property which allows music and dance to be so all-consuming and primal.

The cortical outputs that lead to actions being taken are probably mostly from layer 5 of cortex, just as with the motor outputs of primary motor cortex (Crick,

1984; Koch and Crick, 1994). As with motor outputs, network control outputs are probably graded or tonic signals that are deliberately modulated over time during the control of a network's operation.

The brain command loop's generation of actions within cortex in response to the world state occurs, in adult mammals, from the top down. Higher-level action token regions (which are by no means confined just to frontal cortex, although many are found there) receiving world state inputs (via antecedent support networks) are frequently honed during each period of wakefulness; yielding an abstract action token. This token is then sent on to other networks (including subcortical nuclei) as a *candidate action*. If this action is vetted then this network typically remains active and other, lower-level networks linked to it in hierarchical abstractor networks are also activated as part of the elaboration process for that action. If those lower-level regions (which might be termed *action resources*) were previously engaged in some other action, these actions are suspended and this new higher-level action takes over. This is the fundamental origin of the universal observation that mammals can, in each action resource area, only do one thing at a time (this *action switching* can suddenly redirect those resources to a new action).

Given the activation of a movement action token on a higher-level action region (e.g., for the action of *throwing a ball*), lower-level regions (using a simple derived network design not discussed here) elaborate this token into a temporal sequence of more specific lower-level actions (e.g., designed to elicit motor command tokens that make up the individual postural goals involved in throwing). These lower-level networks can, at each lower level (often there are two or three intermediate levels involved between the abstract action token and the outputs down the spinal cord) also take moment-by-moment world-state inputs (e.g., proprioceptive and somatosensory feedback) and associatively instantiate the tokens of the elaborated action to yield a final action token stream that is adapted to the specifics of conditions at each moment during the performance (learning this multitude of instantiations at each hierarchical level is probably why rehearsal motor skill learning is so arduous). We have demonstrated elaboration and instantiation in simple experiments, but a system capable of competent movement is a long way off.

4.10   Testing this Theory

Scientific theories must be testable. This theory claims that thalamocortical information processing is explained by thousands of interlinked associative memory neural networks of two basic types (feature attractors and antecedent support networks) along with many special-purpose *derived* neural networks (such as the hierarchical abstractor) constructed out of them. Further, it postulates that each of these networks is independently and deliberately activated by a brain command loop that also causes movement (with each network being controlled roughly analogously to a muscle). As with a muscle, each network has a non-negative *activation vigor* that is varied as a function of time. Thalamocortical networks, like muscles, are probably also sometimes afforded continuous *postural* activation input that allows them to process information in a self-continuing string of (rhythmic?) episodes for as long as the input remains present. *Thinking* is nothing more than the self-continuing process of cortical regions producing, during each moment of the *brain command loop*, associative outputs representing *candidate actions* in response to the current (internal and external) *world state*. The candidate commands proceed immediately to subcortical structures (e.g., basal ganglia and various brainstem nuclei) where they are *vetted*. Those actions that successfully pass through vetting are immediately *executed*.

Testing this theory can begin at the bottom or at the top. At the bottom there are the feature attractors. The input of the same stimuli (a dynamic visual stimulus fixed in position on the retina or a brief sound passage supplied directly to the ear canal) embedded in a random ongoing sequence should reliably activate a relatively small collection of standardized tokens within a region. Of course, small changes in the input will alter the region's response, but over a large ensemble of presentations the range of representations should be statistically stable (this will also compensate for the fact that, on any particular trial, only a subset of the neurons of a token actually respond). Thus, a particular token neuron should become active a fixed fraction of the time over ensembles of trials (during which that token is active) carried out on different occasions. For different stimuli other neurons (participating in other tokens) should have this same property. A problem is to figure out which neurons within a region are actually functioning as token neurons; and not acting as transponder neurons for tokens of nearby regions (finding transponder neurons would itself be interesting). So, the test would be to find several neurons which reliably respond to a particular stimulus and then show that these appear together (with reasonable probability) as elements of a standardized token for that stimulus. It would be particularly helpful if the nature of the "active" state could be discovered as part of these investigations.

At the top end, it may be possible to show that activation of a particular feature attractor network (as measured, for example, using the above-suggested protocol) is correlated with a particular thought command input to thalamus or cortex that can separately monitored (perhaps with cortical or thalamic field potential monitoring electrodes or by using fMRI). This would provide concrete evidence of a thalamocortical network activation signal analogous to a motor neuron input to muscle.

In-vivo experiments with antecedent support and hierarchical abstractor networks are probably years away due to the experimental difficulties involved. In the mean time, another approach is building and experimenting with computer-implemented architectures built out of simplified versions of the three thalamocortical network types. The Appendices of this chapter describe such experiments.

4.11   Acknowledgements

The author is grateful to many colleagues, and students for discussions, computer experiments and other contributions to this work. Particular thanks go to colleagues

Robert Means, Kate Mark, David Busby, Rion Snow, and Syrus Nemat-Nasser. The reviews of this chapter by Means and Snow and Todd Gutschow are also appreciated. The support of this research by HNC, ONR, and DARPA is gratefully acknowledged.

Appendix A: Sketch of an Analysis of the Simplified Feature Attractor Associative Memory Neural Network

The set of paired standardized tokens of a simplified feature attractor network (see Section 4.5 and Figure 4.2) can be represented by $\{(\mathbf{x}_k, \mathbf{y}_k)\}_{k=1}^{L}$, where each $\mathbf{x}_k$ and $\mathbf{y}_k$ vector is an element of $\{0,1\}^N \subset \mathbb{R}^N$ having the component corresponding (in some arbitrary but fixed enumeration of the N neurons of that region) to each active neuron set to 1 and all other components set to 0. Each such vector is assumed to have been chosen independently and uniformly at random from among the set of all such vectors possessing exactly M components with value 1 (where $M << N$). Thus, each $\mathbf{x}_k$ and $\mathbf{y}_k$ vector has M components equal to 1 and $N - M$ components equal to 0. [For computer experiments it is often sufficient to simply choose each token component independently and randomly to be 1 with probability M/N and 0 otherwise.]

The set of $\mathbf{x}_k$ vectors and the set of $\mathbf{y}_k$ vectors have, with very high probability, large Hamming distances between the closest distinct pair of vectors within them (for typical biological values of the constants N, M, and L the closest pair of these vectors will overlap by less than 10% of M). Almost all properties of feature attractor networks are statistical and probabilistic in nature. It is easy to construct examples where these networks will fail to function properly. However, the probability of every seeing such a network occurring at random in the real world is negligibly small. Worrying about one of these networks failing is like worrying that all of the molecules of air in the room you are sitting in will, by statistical accident, suddenly rush up into one corner of the ceiling; leaving you to explode from internal pressure in vacuuo. Both could happen, but such improbable events are unworthy of concern. Put another way; you will never look at a car and see your cat.

Now consider what happens when an initial input token $\mathbf{x}$ appears and the feature attractor is applied to it. To see how the analysis goes, assume that $\mathbf{x}$ is equal to $\mathbf{x}_k + \mathbf{d}$, where the *distortion vector* $\mathbf{d}$ has E uniformly randomly chosen components equal to -1 (these are all in coordinates where $\mathbf{x}_k$ is equal to 1) and E uniformly randomly chosen components equal to $+1$ (all in coordinates where $\mathbf{x}_k$ is equal to zero). Thus, $\mathbf{x}$ is a distorted version of $\mathbf{x}_k$ having E of the active neurons of $\mathbf{x}_k$ made inactive and E other neurons not participating in $\mathbf{x}_k$ made active. The value of E varies from 0 to $\lfloor M/2 \rfloor$. Note that since the standardized tokens can be assumed to be far apart, if E is small enough the token $\mathbf{x}$ will be closer to $\mathbf{x}_k$ than to any of the other standardized tokens. An analysis of how the feature attractor will respond to input $\mathbf{x}$ is now sketched.

Given the associative connections established between the C and T regions for the above lexicons (as described in Section 4.5), we can express the responses (total number of inputs received via connections from active neurons on region C – termed the *input intensity* I) of neurons of the T region to the expression of token $\mathbf{x}$ on C as follows. For a particular T region neuron which is part of the token $\mathbf{y}_k$:

$$I = \sum_i \mathbf{x}_i \cdot (\mathbf{x}_k + \mathbf{d})$$

$$= \mathbf{x}_k \cdot \mathbf{x}_k + \sum_{i \neq k} \mathbf{x}_i \cdot \mathbf{x}_k + \sum_i \mathbf{x}_i \cdot \mathbf{d} \qquad (A1)$$

$$= M + \sum_{i \neq k} \mathbf{x}_i \cdot \mathbf{x}_k + \sum_i \mathbf{x}_i \cdot \mathbf{d} \qquad ,$$

and for a particular T region neuron which is not part of the token $\mathbf{y}_k$:

$$I = \sum_j \mathbf{x}_j \cdot (\mathbf{x}_k + \mathbf{d}) = \sum_j \mathbf{x}_j \cdot \mathbf{x}_k + \sum_j \mathbf{x}_j \cdot \mathbf{d} \qquad (A2)$$

where the index i goes over those tokens in which the particular $\mathbf{y}_k$ neuron participates and the index j goes over those tokens in which the non-$\mathbf{y}_k$ token neuron participates.

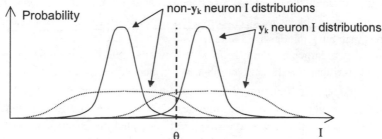

Figure 4.8. When E is near zero (much smaller than M) the distributions of both the $\mathbf{y}_k$ and non-$\mathbf{y}_k$ neuron's I values are relatively narrow (solid lines). Under this circumstance the T region *activation threshold* θ (the cutoff threshold value corresponding to the criterion that exactly M neurons – those with the highest I values – be active on the T region) produces a token very close to $\mathbf{y}_k$ (which, in accordance with Willshaw's theory [Willshaw et al., 1969], when used as the recall key for the C region on the upward part of the feature attractor cycle causes accurate recall of the desired $\mathbf{x}_k$ token). When E becomes too large (close to M) the I distributions widen (dotted lines) and recall causes a massive activation of non-$\mathbf{y}_k$ neurons on the T region. In the biological case – where there is just a threshold, no M quota – this leads to immediate inhibition of the T region because of massive activation of the inhibitory NRT neurons. This is followed by the "resetting" of the network with the null token expressed on the C region (this is how the null token results when the feature attractor input x is too far away from any $\mathbf{x}_k$). An important point about this analysis is that the magnitude of E where this breakdown occurs is (for statistically likely circumstances) very large (an impressive fraction of M). This makes the feature attractor a very robust system for expressing an arbitrary initial input token in terms of a standard token. A similar analysis shows that even significant random neuron death does not effect the network's behavior much either.

The terms in each of the two sums on the right sides of equations A1 and A2 are essentially independent and (within each sum) identically distributed. For typical values of the constants we can apply the central limit theorem to these and approximate each sum very accurately by a Gaussian. Since the sum of two Gaussian random variables is itself a Gaussian random variable, it turns out that the I values can be treated as Gaussian random variables.

As shown in Figure 4.8, in the case of small E both of the I distributions have small standard deviations (relative to their mean values). The difference between the means of these distributions is roughly M. As E increases these standard deviations both increase (because the second sum goes up similarly for both). The difference between the means remains approximately M. This qualitative situation is illustrated in Figure 4.8.

If E is near zero then the activation threshold value θ corresponding to exactly M activated neurons will be such that a large majority of the neurons belonging to token y_k will have their I values above θ. Thus, only a small number of the $N - M$ neurons that are not part of y_k would become active (these would be any far out on the right tail of their I distribution). The net result is activation on the T region of the token y_k, with at most a few deleted and a few added erroneous neurons. When this token is used for recall on the C region (again using the activation threshold that gives exactly M active neurons) the exact token x_k becomes active. So, in one "down and up" cycle the initial input token has been converted by the feature attractor network into the nearest matching standardized token.

As E gets larger (moving from zero towards $\lfloor M/2 \rfloor$) everything keeps working until, at a critical E value, the number of erroneously-activated non-y_k neurons suddenly explodes. The function of the feature attractor network then breaks down. This breakdown might seem very un-biological. However, evoked EEG potential studies have for decades demonstrated a strange effect called the *contingent negative variation* (and other various *late potentials*) that might well an instance of this exact phenomenon (in these experiments the feature attractor is set up in advance – by means of the prior input of an *expectation*, a matter not discussed here – to greatly limit the subset of tokens which may be activated; the non-matching of all of these candidate tokens then causes the above explosive reaction). The sudden mass triggering of thalamic glomeruli probably has the immediate effect of causing the thalamic reticular nucleus (NRT) to strongly and immediately inhibit all of the neurons of the T region. This process probably "erases" the feature attractor (expressing the null token on the C region of the network); preparing it for further use after a short delay. Note that this would happen only on a *second* cycle of the feature attractor (a cycle that doesn't occur in this simplified model): thus explaining why the observed potentials are "late". This further suggests that feature attractor operation may often be "ongoing" (i.e., set up as a sustained rhythmic process that does not require minute control by the brain command loop – see Section 4.5).

So, when the initial C region input token x is *valid* (in the sense of the previous paragraph) a thalamocortical feature attractor reliably converts it to the nearest C region standardized token x_k. When x is not valid the feature attractor produces a null token output and automatically resets itself.

Appendix B: Experiments with a Simplified Antecedent Support Associative Memory Neural Network

This appendix presents some experiments with a simplified antecedent support network implemented using nodes instead of neurons; as described above at the end of Appendix A. The experiments of this appendix and the next will utilize an extensive English text corpus (available from a variety of sources: e.g., the Linguistic Data Consortium and Project Gutenberg). This corpus will be assumed to be diverse (news stories, novels, encyclopedias, etc.) and large (at least many tens of millions of words). The text experiments in my laboratory cited in this chapter have utilized a diverse billion word English corpus.

In these experiments, each cortical region will utilize a lexicon of the 5,000 most common words in your corpus (convert all letters to lower case: retaining and using capitals requires a more sophisticated system than will be described here). Do a word appearance count for all unique words in your corpus and then take the 5,000 top words as your fixed lexicon (numbered from 1 to 5,000). Once you have your lexicon, make another pass through your corpus and replace each word not in this lexicon with word identifier 0. During your use of the lexicon for training you will be extracting *valid strings* of successive contiguous words of variable length. A *valid string* is defined as one which does not contain a period, comma, colon or semicolon within it (i.e., all the words come from the same *clause* of the same sentence), and does not contain a 0 word within it. For each of the training processes used in this and the next appendix, we will be extracting valid strings of a particular stated fixed length; beginning with the first word of the corpus and marching one word at a time to the end of the corpus (viewed as one huge string of text). Each string of words of the desired length is analyzed to see if it is valid; and, if it is, then it is extracted and used for training. If it is not a valid string, then we skip it. In either case, we then move the first word of the candidate string over one word and test that next string. Training stops when we complete the pass through the corpus. A node-based feature attractor will be needed for analyzing the answers produced by the experiments.

The first experiment is illustrated in Figure 4.9 in terms of neurons (but you should implement it in terms of nodes). It consists of two cortical regions, each with its own feature attractor network (only the feature attractor of cortical region β will be used here). The unidirectional fascicle of links from tokens of cortical region α to tokens of cortical region β, along with these two regions themselves, constitute the antecedent support network of this architecture.

The training of an antecedent support network implemented using nodes instead of neurons is very simple. The training process involves keeping track (for example, by means of a 5,000 x 5,000 matrix for each antecedent support network) of the count c(i,j) of how many times token i appeared on the source region of the network at the same time token j appeared on its target region, during the entire training process. Given these counts, the mutual significance s(j|i) of token j in the context of token i can be approximated as:

$$s(j|i) = \frac{\dfrac{c(i,j)}{T}}{\dfrac{c(i)}{T}\ \dfrac{c(j)}{T}} \tag{B1}$$

where T is the total number of valid strings used during training and where c(i) and c(j) are the total number of times source token i and target token j, respectively, were active during the training process. These latter quantities can be derived by the formulas:

$$c(i) = \sum_{j=1}^{L} c(i,j) \quad \text{and} \quad c(j) = \sum_{i=1}^{L} c(i,j) \ . \tag{B2}$$

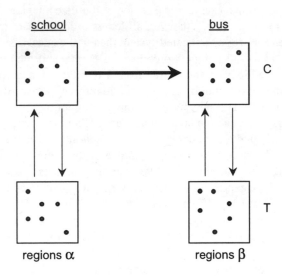

Figure 4.9. A neural network architecture employing two cortical regions and one antecedent support network linking them. This architecture (implemented using nodes as described in the text) is trained on two-word-long valid strings from a large comprehensive text corpus (the feature attractors – of which only the right-hand one is used here – are equipped with the lexicon of the 5,000 most common words of the training corpus). The antecedent support network (its fascicle of links shown in bold) unidirectionally connects cortical region α to cortical region β. After training, this architecture allows exploration of the intuitive character of antecedent support networks and the knowledge they accumulate.

Note that the formula for s(j|i) assumes that the probability of chance co-occurrence of tokens i and j is equal to the product of their a priori probabilities. This is not a completely adequate model of chance word co-occurrences in English, but it will work well enough for the purposes of the experiments of this chapter. One of the main errors this approximate formula for s(j|i) is that the significances for frequent tokens i turn out to be too small.

Once the mutual significances of the token pairs of an antecedent support network have been calculated, then it is time to decide which source and target tokens should be linked. This is done by finding those pairs for which s(j|i) is "significantly greater than 1.0." To do this, pick a significance cutoff threshold (e.g., 2.0 or 3.0 – except for very frequent tokens like "the" and "a," where a threshold of perhaps 0.8 might be used to compensate for the errors in the s(j|i) formula) and discard all but those token pairs which pass this test (i.e., pairs which do not appear together two or three times more often than they would by random chance).

As an aside, this experiment points out how much data must be kept around during learning (which, in thalamocortex, goes on forever) so that link decisions can be made. With a million antecedent support networks in thalamocortex the required "scratchpad memory," is extremely large. This is the beauty of having each transponder neuron synapse autonomously making this decision: the necessary history information can be accumulated by invoking a simple local learning law. Only if the linkage criterion is met does the involve synapse begin to strengthen above baseline – which then automatically causes the synapse (and its paired post-synaptic machinery within the target token neuron to which it is attached) to learn the antecedent support level for that token pair. With 20,000 such synapses for each of the billions of pyramidal neurons in cortex there is more than enough hardware available to maintain and evaluate the vast number of significance statistics involved. Yet only those rare synapses which are actually "used" are "consumed" by this process.

Once the links to be formed have been identified by using the above significance test, the strength of each link (from source token i to target token j) is set to p(i|j), which is calculated using the formula:

$$p(i|j) = \frac{\dfrac{c(i,j)}{T}}{\dfrac{c(j)}{T}} . \tag{B3}$$

Each antecedent support network to be trained must have all of these steps carried out for it (separately from all other such networks being trained). Once these procedures are finished, the network(s) are ready for use.

The first experiment with the trained network of Figure 4.9 (implemented using nodes as described above) is to use it to explore the nature of antecedent support network links. Start by selecting a list of 20 *test words* that you find "interesting" (e.g., united, bird, oil, side, etc.). Before proceeding further, for each of your test words, make a list of ten words that immediately come to mind as words that would "logically follow" as the next word after each test word.

Begin the experiment by expressing the token for each test word (one at a time) on the α cortical region of the network. Then, for each word, apply the trained antecedent support network linking source region α to target region β and make a list of the 10 most excited nodes on region β. Compare these lists with your lists. Eerie, huh?

For example, with our system, here are some of the lists we get (the test word is listed first, followed by the top ten region β words): (**telephone**: **interviews, conversations, conversation, calls, poles, answering, monopoly, lines, booth, numbers**), (**school**: **districts, graduates, administrators, seniors, superintendent, prayer, graduation, teacher, district, uniforms**), (**merit**: **scholarship, raises, scholars, scholar, award, consideration, selection, systems, awards, cigarettes**), (**submit**: **bids, proposals, detailed, applications, recommendations, questions, nominations, written, designs, undated**). Notice that there are no "klunkers." That's because this is how thalamocortex works.

Another interesting experiment is to use a row of five cortical word regions, with the center region linked to and from each of the other regions by an antecedent support network. Train this network on all valid strings of five words in your corpus. Then express one of your test words on the middle region. Go out to each of the other four regions and excite their nodes. Let the *three* most excited nodes (assuming there are any receiving links) on each of these four regions become active (yes, feature attractors can sometimes be used to yield a group of tokens, not just one, by running at a reduced threshold). Then let each of these twelve (or fewer) nodes send their output back to the center region via the reciprocal networks. List the three most excited nodes on the center region. Two of these answers are often words that are synonymous (have the roughly the same meaning and usage) with the word you started with (which is, itself, usually retrieved); aren't they? Hmmm.

Actually this latter experiment should properly be carried out an *ANY* process, but that is too complicated for this introductory overview. The results of this second experiment should simply be used as food for thought.

Hopefully, experiments of this type will give you intuitive insight into the nature of antecedent support learning. This is the only learning that goes on in thalamocortex. Antecedent support knowledge bases contain everything we know.

Appendix C: An Experiment with Consensus Building

In this section an experiment is described that provides an example of how consensus building can be used to bring a collection of assumed facts and knowledge stored in multiple antecedent support neural networks to bear on a question to be answered. The first step in the experiment is to build and train (using the methods of Appendix B) the architecture shown in Figure 4.10.

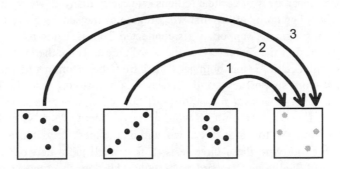

Figure 4.10. Consensus building experiment. Using the methods of Appendix B, three separate antecedent support networks are trained linking the four successive cortical word regions shown (their feature attractor thalamic regions are not shown). These are then used to implement a consensus building process that combines these three knowledge bases to answer the question: given an arbitrary three-word sequence (expressed as tokens on the leftmost three regions); what, logically, should the fourth word be? For example, if the first three words (the assumed facts of the consensus building process) are **down the garden**, consensus building is then carried out on the fourth (answer) region to yield an answer token, which turns out to be (in our system), **path**. This four-word sequence actually occurred in our training corpus. If we then blank out the token for the word path and repeat the thought process, the next word obtained is *lane*; which was not ever seen (with the three first words) in the training corpus (the italics on the answer indicate that it is *novel* in this sense). Consensus building is the universal information processing operation of thalamocortex.

Once your architecture is trained, then carry out consensus building (using the node honing procedure described in Appendix B) using various collections of assumed facts (the first three words, as explained in the caption of Figure 4.10). What you will quickly realize is that, even with only three knowledge bases (we probably use several tens of knowledge bases, context, and iterative testing and refinement of our candidate answers when we answer such phrase-completion questions), this system is amazing. Here are some typical examples from our system: **driving at top _speed_**, **all of a _sudden_**, **traveling around the _country_**, and **down the garden _path_**. When we blank each of these answers and obtain a second answer the following words are obtained, respectively: **_speeds_**, **_common_**, **_village_**, and **_lane_**. Again, these results are typical. Downright eerie; unless, of course, this theory is correct.

An interesting aspect of consensus building is that at some point in the process of obtaining successive answers there will be no more logically viable answers (i.e., the null token or node appears as the answer). Thus, consensus building is automatically self-limiting: when the process runs out of answers it stops. To demonstrate this in your own thalamocortex, make a list of all words that would logically follow **wonderful to have**. Then try out your architecture and see what it produces. Hmmm.

Having trained the three antecedent support networks of the Figure 4.10 (I will refer to them by their labels there: 1, 2, and 3), you can try another, totally different experiment with them. The goal of this experiment is to detect phrases in novel (i.e., not in the training corpus) text. Imagine the words of a sentence expressed, in sequence on a linear array of cortical regions (we use as many as we need to hold the sentence). Then, beginning at the first word region of the sentence, we test to see if the second word logically supports it (is connected from it) using network 1. If not, then the first word of the sentence is a one-word phrase. If so, then we test the third word of the sentence to see that both network 1 from the second word and network 2 from the first word support it (yes, this is essentially just consensus building being used to test a supplied answer for validity). If both of these words support the third word, then we move onto the fourth word and apply networks 1, 2, and 3 and our consensus building test to words three, two, and one, respectively. It turns out that we can just keep going with these three tests (a four-word jump network doesn't add very much) until finally the last word tested fails. This then constitutes a phrase.

This phrase detector system can be used with a hierarchical abstractor to automatically build a higher-level region with a lexicon of all English phrases (for a 5,000 word vocabulary there are over a hundred thousand valid phrases). We can then go up to this higher abstraction level and begin to learn the relationships between adjacent phrases by means of antecedent support networks linking these regions. We can also link phrase regions to word regions; and even keep moving up the hierarchy to sentence regions and so forth. All of this works. The results are mind-blowing. Well, actually, they are probably just mind-mimicking.

Finally, as you carry out these experiments, ponder the questions of What kind of knowledge is being learned? and How is that knowledge being used? The answers to these questions will help illustrate the utter novelty and enormous power of this approach to information processing. Fuzzy knowledge, no rules. All available knowledge that is directly applicable to directly answering a poseable question is immediately and automatically available. Novel situations are handled; because the stored knowledge is universally applicable to all questions that the system is prepared to answer. These attributes (and many, many others possessed by these systems) are those which AI and neuroscience have sought, with absolutely no success, in over four decades of attempts to build models of thalamocortical information processing.

In conclusion, it seems plausible that the theory presented here is the long-sought answer to the basic question of how the mammalian thalamocortex works. It may also be the royal road to artificial intelligence. Wonderful days lie ahead for us all.

References

Abeles, M. (1991) Corticonics. Cambridge, UK: Cambridge Univ. Press.

Amari, S. (1989) Characteristics of sparsely encoded associative memory. Neural Networks 2: 451–457.

Amari, S. (1974) A method of statistical neurodynamics. Biological Cybernetics 14: 201–215.

Amari, S. (1972) Learning patterns and pattern sequences by self-organizing nets of threshold elements. IEEE Trans. Comput. C-21: 1197–1206.

Anderson, J.A. (1968) A memory storage model utilizing spatial correlation functions. Kybernetik 5: 113–119.

Anderson, J.A. (1972) A simple neural network generating an interactive memory. Mathematical Biosciences 14: 197–220.

Anderson, J.A., Silverstein, J.W., Ritz, S.A., Jones, R.S. (1977) Distinctive features, categorical perception, and probability learning: Some applications of a neural model. Psych. Rev. 84: 413–451.

Bender, E.A. (1996) Mathematical Methods in Artificial Intelligence. Los Alamitos, CA: IEEE Computer Society Press.

Bennett, C.H., Gacs, P., Li, M., Vitanyi, P.M.B., Zurek, W.H. (1998) Information distance. IEEE Trans. Infor. Th. 44: 1407–1423.

Borisyuk, R., Borisyuk, G., Kazanovich, Y. (1998) Synchronization of neural activity and information processing. Behavioral and Brain Sciences 21: 833–844.

Braitenberg V., Schuz, A. (1998) Cortex: Statistics and Geometry of Neuronal Connectivity, Second Edition. Berlin: Springer-Verlag.

Carpenter, G.A., Grossberg, S. (Eds.) (1991) Pattern Recognition by Self-Organizing Neural Networks. Cambridge, MA: MIT Press.

Carpenter, G.A., Grossberg, S. (1988) Adaptive Resonance. In: Grossberg, S. (Ed.) Neural Networks and Natural Intelligence. Cambridge, MA: MIT Press, 251–315.

Chover, J. (1996) Neural correlation via random connections. Neural Computation 8: 1711–1729.

Cohen, M.A., Grossberg, S. (1983) Absolute stability of global pattern formation and parallel memory storage by competitive neural networks. IEEE Trans. Sys. Man and Cyber. 13: 815–826.

Cowan, W.M., Sudhof, T.C., Stevens, C.F. (Eds.) (2001) Synapses. Baltimore, MD: Johns Hopkins Univ. Press.

Crick, F.H.C. (1984) Function of the thalamic reticular complex: The searchlight hypothesis. Proc. Nat. Acad. Sci. 81: 4586–4590.

Fain, G.L. (1999) Molecular and Cellular Physiology of Neurons. Cambridge, MA: Harvard Univ. Press.

Fries, P., Reynolds, J.H., Rorie, A.E., Desimone, R. (2001) Modulation of oscillatory neuronal synchronization by selective visual attention. Science 291: 1560–1563.

Fukushima, K. (1975) Cognitron: A self-organizing multilayered neural network. Biological Cybernetics 20: 121–136.

Fukushima, K., Miyake, S. (1978) A self-organizing neural network with a function of associative memory: Feedback-type Cognitron. Biological Cybernetics 28: 201–208.

Gerstein, G.L. (2001) Neural assemblies: Technical issues, analysis, and modeling. Neural Networks 14: 589–598.

Grossberg, S., Williamson, J.R. (2001) A neural model of how horizontal and inter-laminar connections of visual cortex develop into adult circuits that carry out perceptual groupings and learning. Cerebral Cortex 11: 37–58.

Grossberg, S. (1999) How does the cerebral cortex work? Learning, attention and grouping by the laminar circuits of visual cortex. Spatial Vision 12: 163–186.

Grossberg, S., Mingolla, E., Ross, W.D. (1997) Visual brain and visual perception: How does the cortex do perceptual grouping? Trends in Neurosciences 20: 106–111.

Grossberg, S. (1997) Cortical dynamics of three-dimensional figure-ground perception of two-dimensional patterns. Psych. Rev. 104: 618–658.

Grossberg, S. (1995) The attentive brain. American Scientist 83: 438–449.

Grossberg, S. (Ed.) (1987) The Adaptive Brain, Volumes I and II. Amsterdam: Elsevier.

Grossberg, S. (Ed.) (1982) Studies of Mind and Brain. Norwell, MA: Kluwer.

Grossberg, S. (1976) Adaptive pattern classification and universal recoding. Biological Cybernetics 23: 121–134.

Grossberg, S. (1968) Some nonlinear networks capable of learning a spatial pattern of arbitrary complexity. Proceedings of the National Academy of Sciences 59: 368–372.

Haines, K., Hecht-Nielsen, R. (1988) A BAM with increase information storage capacity. Proceedings, 1988 International Conf. on Neural Networks, Piscataway NJ: IEEE Press, I-181–I-190.

Hall, Z.W. (1992) Molecular Neurobiology. Sunderland, MA: Sinauer.

Hebb, D. (1949) The Organization of Behavior. New York: Wiley.

Herculano-Houzel, S., Munk, M.H.J., Neuenschwander, S., Singer, W. (1999) Precisely synchronized oscillatory firing patterns require electroencephalographic activation. J. Neurosci. 19: 3992–4010.

Hopfield, J.J. (1982) Neural networks and physical systems with emergent collective computational abilities. Proc. Natl. Acad. Sci. 79: 2554–2558.

Izhikevich, E.M. (2001) Resonate-and-fire neurons. Neural Networks 14: 883–894.

Jones, E.G. (1985) The Thalamus. New York: Plenum Press.

Kainen, P.C., Kůrková, V. (1993) On quasiorthogonal dimension of euclidean space. Applied Math Lett. 6: 7–10.

Koch, C., Crick, F.H.C. (1994) Some further ideas regarding the neuronal basis of awareness, In: Koch, C., Davis, J. (Eds.) Large-Scale Neuronal Theories of the Brain. Cambridge, MA: MIT Press.

Kohonen, T. (1995) Self-Organizing Maps. Berlin: Springer-Verlag.

Kohonen, T. (1984) Self-Organization and Associative Memory. Berlin: Springer-Verlag.

Kohonen, T. (1972) Correlation matrix memories. IEEE Transactions on Computers C21: 353–359.

Konig, P., Engel, A.K., Roelfsema, P.R., Singer, W. (1995) How precise is neuronal synchronization. Neural Computation 7: 469–485.

Kosko, B. (1988) Bidirectional associative memories. IEEE Trans. On Systems, Man, and Cybernetics SMC-18, 49-60.

Kryukov, V.I., Borisyuk, G.N., Borisyuk, R.M., Kirillov, A.B., Kovalenko, E.I. (1990) Metastable and unstable states in the brain. In: R.L. Dobrushin, V.I. Kryukov, A.L. Toom (Eds.) Stochastic Cellular Systems. Manchester, UK: Manchester Univ. Press.

Ledoux, M. (2001) The Concentration of Measure Phenomenon. Providence, RI: American Mathematical Society.

Miller, G. A. (1996) The Science of Words. New York: Scientific American Library.

Mountcastle, V.B. (1998) Perceptual Neuroscience: The Cerebral Cortex. Cambridge, MA: Harvard Univ. Press.

Nakano, K. (1972) Associatron - a model of associative memory. IEEE Transactions on Systems, Man, and Cybernetics SMC-2: 380–388.

Nicholls, J.G., Martin, A.R., Wallace, B.G., Fuchs, P.A. (2001) From Neuron to Brain, Fourth Edition. Sunderland, MA: Sinauer.

Nicolelis, M.A.L., Baccala, L.A., Lin, R.C., Chapin, J.K. (1995) Sensorimotor encoding by synchronous neural ensemble activity at multiple levels of the somatosensory system. Science 268: 1353–1358.

Nilsson, N.J. (1998) Artificial Intelligence: A New Synthesis. San Francisco: Morgan Freeman Publishers.

Nilsson, N.J. (1965) Learning Machines. New York: McGraw-Hill.

Nolte, J. (1999) The Human Brain, Fourth Edition. St. Louis, MO: Mosby.

Oja, E. (1980) On the convergence of an associative learning algorithm in the presence of noise. International Journal of Systems Science 11: 629–640.

Palm, G. (1980) On associative memory. Biol. Cybern. 36: 19–31.

Sagi, B., Nemat-Nasser, S.C., Kerr, R., Hayek, R., Downing, C., Hecht-Nielsen, R. (2001) A biologically motivated solution to the cocktail party problem. Neural Computation 13: 1575-1602.

Schneider, W.X., Owen, A.M., Duncan, J. (Eds.) (2000) Executive Control and the Frontal Lobe. Berlin: Springer-Verlag.

Sejnowski, T.J., Destexhe, A. (2000) Why do we sleep? Brain Research 886: 208–223.

Sherman, S.M., Guillery, R.W. (2001) Exploring the Thalamus. San Diego, CA: Academic Press.

Shin, J. (2001) Adaptation in spiking neurons based on the noise shaping neural coding hypothesis. Neural Networks 14: 907–919.

Sommer, F.T., Palm, G. (1999) Improved bidirectional retrieval of sparse patterns stored by Hebbian learning. Neural Networks 12: 281–297.

Steinbuch, K. (1961a) Automat und Mensch. Heidelberg: Springer-Verlag.

Steinbuch, K. (1963) Automat und Mensch, Second Edition. Heidelberg: Springer-Verlag.

Steinbuch, K. (1965) Automat und Mensch, Third Edition. Heidelberg: Springer-Verlag.

Steinbuch, K. (1961b) Die lernmatrix. Kybernetik 1: 36–45.

Steinbuch, K., Piske, U.A.W. (1963) Learning matrices and their applications. IEEE Transactions on Electronic Computers December: 846–862.

Steinbuch, K., Widrow, B. (1965) A critical comparison of two kinds of adaptive classification networks. IEEE Transactions on Electronic Computers October: 737–740.

Steward, O. (2000) Functional Neuroscience. New York: Springer-Verlag.

Stryker, M.P. (2001) Drums keep pounding a rhythm in the brain. Science 291: 1506–1507.

Thorpe, S., Delorme, A., Van Rullen, R. (2001) Spike-based strategies for rapid processing. Neural Networks 14: 715–725.

Ukhtomsky, A.A. (1966) Dominanta (in Russian). Leningrad: USSR Academy of Sciences.

von der Malsburg, C. (1981) The correlation theory of brain function. Internal Report 81-2, Max-Planck-Inst. for Biophysical Chemistry.

Widrow, B., Hoff, M.E. (1960) Adaptive switching circuits. 1960 IRE WESCON Convention Record, New York: Institute of Radio Engineers, 96-104.

Willshaw, D.J., Buneman, O.P., Longuet-Higgins, H.C. (1969) Non-holographic associative memory. Nature 222: 960–962.

Zadeh, L.A. (1965) Fuzzy sets. Information and Control, 8, 338-353.

Zador, P. (1963) Development and Evaluation of Procedures for Quantizing Multivariate Distributions, PhD Dissertation, Palo Alto, CA: Stanford University.

Chapter 5

Elementary Principles of Nonlinear Synaptic Transmission

Henry Markram

5.1 Abstract

The efficacy of synaptic transmission changes dynamically from one action potential to the next. This dynamic transmission results in nonlinear input-output functions at synapses. The lack of quantitative experimental data on nonlinear synaptic transfer has hindered the incorporation of these synapses into comprehensive theories of brain function. Recent experiments, however, between specific types of neurons in the neocortex provide quantitative data and novel insight into synaptic operations. Based on these data, elementary principles of nonlinear synaptic transmission are derived here and are invoked to propose a general theory of information processing, learning, and memory in recurrent neural networks. It is proposed that temporal features of the environment are processed by nonlinear synaptic transmission and by recurrent interactions between neurons and that recurrent circuitry serves to generate a continual neural representation of past information that is embedded in single neurons and in the connectivity structure of the network. In this theory, all memories are potentially instantly available from the current state of activity and a major component of learning is where the microcircuit learns about itself in order to align the functions of each of its components (molecules, synapses and neurons).

5.2 Introduction

It is not over when an impulse flashes across a synapse and onto its destination.
It leaves behind ripples in the state of the system.

Ralph Gerard, 1949

The ripples that Gerard refers to are multiple time-scaled biophysical and biochemical postsynaptic effects that persist after the transmission event, but "ripples" are also left presynaptically. "Ripples" are unavoidable and provide an immensely powerful platform for temporal integration. A striking feature of synaptic transmission, observed in numerous organisms and multiple brain regions, is that subsequent action potentials (APs) in a train are not transmitted equally (Figure 5.1). This is because each AP leaves behind "ripples" that influence subsequent transmission. The result is that the input-output function of synapses is not a linear one. For the sake of simplicity, most neural network theories have "linearized" synaptic transmission (Appendix A, Sherrington's Leap) and the significance of this universal feature of synapses has therefore not been fully developed (Appendix B, Functional Significance).

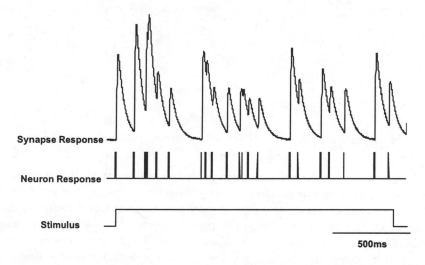

Figure 5.1. Typical synaptic response of a depressing synaptic connection to an irregular train of APs emitted by a presynaptic neuron (simulation).

The biophysical mechanism of nonlinear synaptic transmission has, on the other hand, received much more attention, but even so, the precise mechanisms of dynamic neurotransmitter release still remain elusive. Indeed, what clearly emerged is that the synapse is one of the most complex submicron devices in the nervous system (see Korn et al., 1986; Zucker, 1989; Pongs et al., 1993; Pevsner and Scheller, 1994; Parnas and Parnas, 1994; Brodin L. and Shupliakov, 1994; Henkel and Almers, 1996; Shupliakov et al., 1997; Murthy and Stevens, 1998; Ales et al., 1998; Thomson, 2000). Neurotransmitter vesicles are found in various stages of readiness as they approach the docking site for release and an extraordinarily complex machinery, consisting of numerous interacting molecules, form a leverage mechanism to induce fusion of the vesicle membrane with the terminal membrane (see Thomson, 2000).

The precision with which these complex molecular interactions occur is awesome; release can occur in as little as 100µs after Ca^{2+} enters the terminal. Vesicle readiness, the specific molecules involved in the lever and specific Ca^{2+} competitor proteins operate to determine the probability that a pulsed influx of Ca^{2+} will trigger release of transmitter. By adjusting the size of the different pools of vesicles, the specific molecules used in the lever, the nature of the Ca^{2+} competitor proteins and the efficacy of Ca^{2+} entry into the terminal, synapses may reset more or less rapidly after each release giving rise to different degrees of synaptic depression to subsequent responses. The speed of this reset determines how efficiently synapses can respond to successive APs. Some synapses also begin to release with low probabilities and then "warm up" (facilitate) as the synapse is stimulated repetitively – this priming effect can be brief or persist for several seconds or minutes.

Each synapse is also under direct modulatory control from many other types of neurons and each type of neuron employs different neurotransmitter systems that act through one of a multitude of intra-terminal pathways to regulate synapses in potentially subtle ways that have not even begun to be explored. The molecular complexity of synapses provides the potential for immense diversity in the types of synapses employed in the nervous system. This seems a massive overkill if synapses are simply to perform an electrochemical transduction that is only subject to increasing and decreasing the efficiency of transduction.

Synapses between individual neurons in the neocortex are characterized by a rich diversity of nonlinearities (see Thomson and Deuchars, 1994, 1997; Markram, 1997; Markram et al., 1998c; Wang et al., 1999). It is difficult to understand how theories of brain function could exclude such dynamic coupling between neurons and even more so, how the current concept of learning and memory could be based on mere linear scaling of the strength of synapses. This intriguing dynamic coupling between neurons triggered a series of experiments aimed at understanding the essential operations performed by a recurrent microcircuit of neurons connected with nonlinear synapses, such as a neocortical column. The approach was to record synaptic connections between pre-selected and specific pre- and postsynaptic neurons (Appendix C, Visual Patch Recordings) and to quantify the synaptic responses to various conditions of stimulation. This quantification enabled construction of a phenomenological model (Tsodyks and Markram, 1997; Markram et al., 1998a; Tsodyks et al., 1998). This model allowed us to integrate experimental and theoretical studies and mathematically explore further the operations performed by synapses. The main conclusion from these studies is that synapses do not simply transmit AP information from one neuron to the next, but rather engage the APs to determine the actual form of the information.

This chapter summarizes these experiments and theoretical work and attempts to provide some framework for incorporating nonlinear synapses into brain theories. An attempt is made to derive reduced elementary principles of nonlinear synaptic transmission. These principles prescribe basic physical constraints for information transmission by nonlinear synapses. In its later part, the physical constraints for storing information, setting parameters and retrieving memories from nonlinear synapses are outlined. Synaptic principles are invoked to propose a general theory of information processing in networks of neurons recurrently interconnected with nonlinear synapses.

5.3   Frequency-dependent Synaptic Transmission

> Under natural conditions synapses are activated by trains of impulses that may
> be of relatively high frequency... It is therefore imperative to study the opera-
> tion of synapses during repetitive activation.
>
> *John Carew Eccles, 1964*

Nonlinear synaptic transmission arises because the efficacy of transmitting one AP
depends on the transmission of earlier APs. Each AP has multiple effects that interact
and decay differentially in time so that the transmission by any one AP is influenced
by the sum of the residual effects of all previous APs. The outcome of the synapse at
any moment in time depends on the current state of "perturbation" of the synaptic
machinery. To describe synaptic transmission as "frequency-dependent", averages
this sensitivity to the precise temporal structure of the input. This average approach
proves to be very useful in generalizing the operations performed by synapses, but
can only provide partial insight into the operations of synapses.

Frequency-dependent transmission is due to a dynamic interplay between the rate
of neurotransmitter release, the rate of recovery from release and the rate of recovery
from facilitation of release. Synaptic transmission is probabilistic and the average
rate of release is therefore the product of the probability of release (p) and the frequency of stimulation (f), $R_r = p \cdot f$.

A refractory period follows each release event and this underlies a basic rule of
"release-dependent depression". In some synapses, a priming or facilitation of release
follows each AP arriving in the synaptic terminal and this underlies a basic rule of
"release-independent facilitation". The refractory period (depression) seems to be
present at all synapses, but to widely different degrees, while the priming mechanism
(facilitation) may be strongly, weakly or not expressed at all.

Probabilistic release of transmitter largely determines the importance of the other
parameters since p provides the range for facilitation and the rate of release determines the degree to which synaptic depression is engaged which in turn constrains
facilitation. The extent and rate of synaptic depression is therefore approximately
proportional to p and the extent and rate of synaptic facilitation is approximately
inversely proportional to p. For example, when p is low then each AP can facilitate p,
but as p grows, the dynamic range for facilitation decreases and depression is more
fully engaged. This analysis is true in terms of actual release of transmitter, but the
functional expression of depression and facilitation also depends on the manner in
which the postsynaptic receptors are engaged. The parameter p is therefore intuitive,
useful and probably valid for analyzing many synapses (since receptors may not
always be limiting), but an equivalent generalizable description is preferred in which
the equivalent parameter for p is referred to as the "utilization of synaptic efficacy"
(u) (Appendix D, Biophysical Basis of Parameters).

Dynamic synaptic transmission is actually much more complex than presented
here. Synaptic depression can have at least two time constants and facilitation can
have as many as 4 time constants (2 for facilitation, 1 for augmentation and 1 for
potentiation; see Magleby, 1987). Synaptic depression may also be engaged when the

release machinery is driven to the furthest point, but just fails, hence yielding release-independent depression. Synaptic depression may also be a variable where repeated stimulation can speed up the rate of recovery of the release machinery. Furthermore, facilitation may result from the sensitization or unmasking of postsynaptic receptors, which would yield "release-dependent facilitation". The proportions of these different mechanisms that are active at different synapses is still not known. Nevertheless, while the complete story of the synaptic dynamics is still unfolding, we can already derive a basic framework for the organization and function of some of the essential principles of nonlinear synaptic transmission.

Characteristic parameters of synapses and synaptic connections are therefore: the sign of the connection, which is determined by the type of neurotransmitter receptors activated; the maximal response produced by a single release event, referred to as the quantal size (q). In the TM model q is the appropriate definition for the absolute strength of a single release site and is referred to as a. The strength of a synaptic connection as a whole, made up of n release sites is $A = q.n$ (Appendix E, Single Connection, Many Synapses). The probability of neurotransmitter release is p and the average fraction of A utilized by APs is U, which under certain conditions is equivalent to average p (Appendix D, Biophysical Basis of Parameters). The time course of recovery from the refractory period of release that underlies synaptic depression is τ_{rec} and the time course of facilitation of U is τ_{facil}. The running variable of the parameter U is u.

Dynamic interaction between U, τ_{rec} and τ_{facil} determines the moment-to-moment response of the synapse to a presynaptic AP train and these three parameters are therefore also referred to as "kinetic parameters".

Synaptic Parameters

q = quantal size
a = absolute strength of a single synapse
A = absolute strength of a synaptic connection
p = probability of neurotransmitter release
U = utilization of synaptic efficacy
u = facilitated U
R = synaptic resources in the recovered state
τ_{rec} = time constant for recovery from depression
τ_{facil} = time constant for recovery from facilitation

In order to derive the parameters of a synaptic connection, a phenomenological model that considers the interplay between U, τ_{rec} and τ_{facil}, is used to fit average responses to trains of presynaptic APs (Appendix F, The Model). For example, at a purely depressing synapse, if U is high then after the first AP, the synapse is refractory and depression is observed for the second response if it occurs within about $3.\tau_{rec}$. The form of the depression therefore carries information about both p and the rate of recovery from depression. By experimentally measuring the recovery from depression the contribution of the rate of recovery from depression can be peeled out and the parameter U can be derived. A is then the response that would be observed if U were 1.

At facilitating synapses, depression is countered by an increase in U and hence the extent to which responses deviate from pure depression can be used to derive τ_{facil}. We classify synapses into two broad categories (Figure 5.2) based on whether a mechanism of facilitation is active or not. Purely depressing synapses lack a facilitatory mechanism while facilitating synapses may appear as depressing synapses if p is high, but these synapses are still distinct from pure depressing synapses. Current studies indicate that several canonical classes and subclasses of synapses exist depending on relative time constants of depression and facilitation and on U (Appendix G, Synaptic Classes).

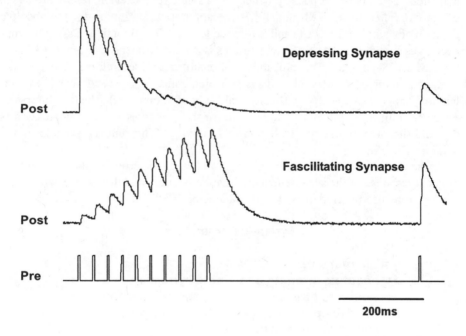

Figure 5.2. Two basic classes of synapses. Depressing synapses are defined as synapses that exhibit no functional facilitation and are different from high probability of release facilitating synapses.

5.4   Nonlinear Synapses Enable Temporal Integration

If APs in a train are randomly spaced (see Softky and Koch, 1993), why are synapses then so sensitive to the precise timing of the APs? Noisy or randomly occurring APs suggest that each AP is essentially independent and that the sum of the information carried by individual APs in a train is equivalent to the information carried by the whole train; in other words, the train does not contain more information than the sum of its components.

For linear "synapses" this is true since each synaptic response is independent of previous synaptic responses and hence, at the synaptic level, the information is proportional to the sum of information carried by each AP. For nonlinear synapses, however, the synaptic response to each AP is dependent on its relative timing with respect to other APs in the train (Figure 5.3). Each AP therefore carries some information about the fact that it occurred as well as some information about the history of APs. The information transmitted by each AP then depends on the residual information transmitted by all proceeding APs. The content of the information transmitted by the train as a whole therefore concerns how each AP is temporally related to each other AP in a train. In other words, nonlinear synaptic transmission is ideal for integrating temporal information.

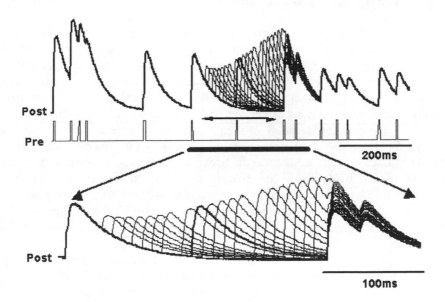

Figure 5.3. Synaptic responses to repeated trials of the same irregular train of APs with a progressive shift in the timing of one AP (at the center of the arrow in the top traces, which are expanded below). The traces below show a time expanded representation of the effect of shifting the timing of the AP. Simulation of average synaptic responses.

To consider nonlinear synapses as decoding the AP train is not strictly correct. The information carried by the AP train is not an absolute form of information, but rather *potential* information since the particular synaptic nonlinearities largely determine the relevant contribution of each AP. On the other hand, synapses also do not contain an absolute form of information since the patterns of APs determine the response of the synapse.

The actualized information transmitted by nonlinear synapses is therefore seen as the result of an *interaction* between a form of explicit potential information carried by the AP train, and a form of implicit potential information contained within the synapse. The first and most elementary principle of the function of nonlinear synaptic transmission is therefore that the perturbed state of synaptic processes resulting from a train of APs represents the temporal integration of the sequence of preceding APs and the response to the current AP represents the readout of this temporal information.

First Principle of the Function of Nonlinear Synapses:

The interaction between the AP train and the synaptic machinery generates a perturbed synaptic state that represents past inputs.

5.5   Temporal Information

Information theory can be employed to assess the information carried by single or trains of APs, but in this chapter, the intuitive basis for such an analysis is developed. If a single synaptic response at a specific moment in time is considered then many different patterns of APs preceding the moment could give rise to the same amplitude response, suggesting that if only a single isolated synaptic response were considered then the precise relative timing of the APs within a train would not be critical to the network. In general, the accuracy of determining the history of activity (i.e. the sequence of preceding APs) from a single response would be poor and would increase as U and τ_{rec} increases since the relevant history is squeezed into the previous AP response. If organisms then process information in neural networks based exclusively on the synaptic response to selected responses in a train then the network cannot rely on the relative timing between APs as an unambiguous source of information, instead information would have to be processed according to some absolute temporal scale.

If two subsequent synaptic responses are considered then the number of AP patterns that could lead up to and produce those responses is less than if only one is considered, but it could still be large (Appendix H, Paired Pulses). The more synaptic responses in a sequence are considered, the fewer the possible AP patterns become that can produce those responses. The information about the precise AP pattern therefore increases as a function of the number of synaptic responses considered in the train. A unique function of nonlinear, as opposed to linear synaptic transmission, is therefore that a sequence of amplitudes can code for a sequence of interspike intervals. If organisms therefore process information using trains of APs (as opposed to isolated APs) then the sequence of synaptic response amplitudes is an unambiguous source of information about the precise nature of AP activity in the network.

Second Principle of the Function of Nonlinear Synapses:

A sequence of interspike intervals can be represented by a synapse in an unambiguous manner by a vector of synaptic response amplitudes.

Action potential trains generate unique synaptic responses provided that sufficient APs in the train are considered. One way to illustrate this uniqueness is to examine the variation in the amplitude of synaptic responses (SRs) evoked during a train of APs as a function of the variation in the interspike interval (ISI) of APs in the train with the same mean frequency. While the coefficient of variation (CV) of SRs evoked by a train could be the same for different sequences of synaptic responses, different CVs nevertheless provide an indication of uniqueness. In general, for nonlinear synapses, the CV of SRs increases as a function of the CV of ISIs (Figure 5.4). Different patterns of APs with the same average frequency of discharge produce a scatter of CVs indicating that each train produces a unique response and therefore potentially carries unique information. The distributions of CVs of SRs generated by different patterns of APs of the same average frequency depends critically on synaptic parameters (Figure 5.4).

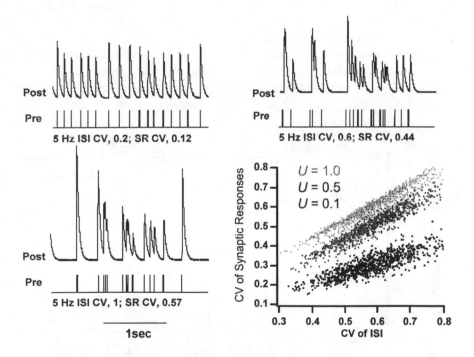

Figure 5.4. The effect of AP irregularity on the coefficient of variation (CV) of synaptic responses (SRs). In the traces, the same average frequency of APs was generated, but the CV of the ISI was increased. In the graph three sets of data are represented. In the lower dots the CV of SR versus CV of ISI is plotted for 500 trials of different realizations of an irregular trains of 10Hz. The U parameter for the lower dots was 0.1. The U parameter for the middle dots was 0.5. The U parameter for the upper dots was 1.0.

5.6   Packaging Temporal Information

An interesting emergent concept, which is essentially opposite in principle to many neural network theories of information representation, is that temporal information is maximal at nonlinear synapses when the synaptic machinery is the furthest from a steady state (away from equilibrium; the highest energy state or the state of maximal perturbation). A train of APs initially perturbs synapses so that the rate of release, rate of depression of release and rate of facilitation of release are unbalanced causing the synaptic response to either increase or decrease for successive APs. This is referred to as a non-equilibrium phase of transmission. Specific APs in a train, that mark the onset of a non-equilibrium phase can be determined by measuring the instantaneous transmission; product of the synaptic response and the instantaneous frequency. For depressing synapses, when the synapse is near equilibrium then short ISIs (high instantaneous frequencies) are correlated with smaller synaptic responses and vice versa. The instantaneous transmission therefore deviates only when an AP is placed within an irregular train such that it tends to drive the synapse towards a new equilibrium. The instantaneous transmission can therefore isolate all "first" APs in a train of APs that are due to a non-stationary (novel) stimulus (Figure 5.5). This measure normalizes AP irregularity that is consistent with the current trajectory of the synapse towards an equilibrium position and emphasizes the onset of events that are driving the synapse towards a new state. For linear "synapses" the rate of transmission would change simply as a function of the irregularity of the AP train.

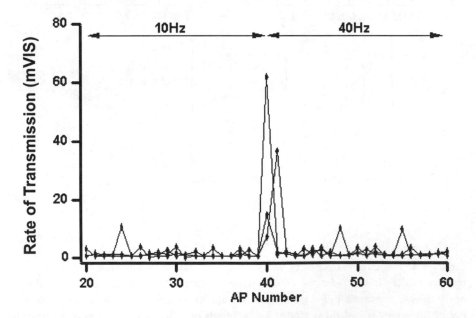

Figure 5.5. Instantaneous transmission. The y-scale represents the product of the synaptic response (SR) and the instantaneous frequency; $SR * 1/ ISI$ of AP_n and AP_{n-1}.

Transmission by nonlinear synapses therefore occurs in a spectrum of equilibrium and non-equilibrium phases. Properties of the equilibrium and non-equilibrium periods are determined by the synaptic properties and the nature of the stimulus. The non-equilibrium period is not only dependent on the frequencies before and after a stimulus, but also on the irregularity of the APs in the train. The effect of AP irregularity on the duration of the non-equilibrium period is approximately proportional to the number of APs required to be certain about the average frequency in a train; the square of the standard deviation of the 1/ISIs divided by the product of the actual mean frequency and some certainty level.

Nonlinear synapses are therefore able to transmit an unambiguous signal to the postsynaptic neuron concerning the structure of the AP train *only* when enough APs have been transmitted. In other words, if one considers a neural network with x nonlinear synapses there are very definite constraints as to the number of APs that must be transmitted by each synapse before the synaptic responses that are generated in the network reflect the settings within the synapse. With too few APs the responses could have been generated by any synapse (equivalent to arbitrary synaptic settings). This physical constraint becomes even more serious when synaptic nonlinearities are dictated by many synaptic parameters (such as multiple time constants), because an even larger number of APs will be required to reach synaptic unambiguity. There is currently no theoretical framework to describe this activity constraint mathematically for neural networks.

As more APs are transmitted by a synapse the temporal information (TI) reaches a maximum. The limit of TI is reached as the synapse approaches equilibrium and the number of APs required to construct a complete and unambiguous representation of the structure of the AP train is therefore essentially the same as that required to drive the synapse to equilibrium or quasi-equilibrium. It is actually also the same constraint for modeling the nonlinear synapse in order to extract the synaptic parameters unambiguously.

The third principle of the function of nonlinear synaptic transmission is therefore that temporal information is packaged in the non-equilibrium phase of the synaptic transmission. This principle provides the basic physical constraints for how many APs and their structure are required for an unambiguous representation of presynaptic activity as a function of the synaptic parameters. Interestingly, this principle may explain why most neuronal discharges are described as Poisson; maximum irregularity is optimal for keeping the synapse in a maximally perturbed state.

Third Principle of the Function of Nonlinear Synapses:

Temporal information is packaged in the non-equilibrium phase of synaptic transmission.

5.7   Size of Temporal Information Packages

If an AP stimulus were stretched out in time then TI would be completely lost. An analogous psychophysical effect could be where a sentence, paragraph, song, poem or even sliding of one's finger along a surface to determine the texture must be carried out within a critical time; a critical associative time. The critical associative time for synapses is the duration of the non-equilibrium period. Several principles can be derived for how synaptic parameters affect the duration of the non-equilibrium period and therefore also the size of the packets of TI. Synapses that approach steady state rapidly are able to transmit an unambiguous signal using as few as two APs (Figure 5.6), but the package contains a minimal amount of TI. An analogy could be the binding of elementary temporal features of a stimulus, such as binding one or two letters in a word or a few harmonics in a sound. The longer the time required to reach steady state or quasi steady state, the more TI can be packaged. The non-equilibrium period of low p synapses can last many seconds (information transmitted by APs much later are still affected by an event), indicating that these synapses can transmit

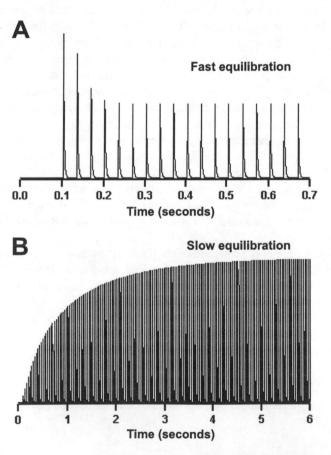

Figure 5.6. Approaching equilibrium in transmission. Simulated current responses evoked by trains of regular APs. (A) Rapidly depressing synapse with high U (0.7) and no facilitation. (B) Low U (0.01) facilitating synapse (τ_{facil}, 2000ms) with fast time constant of recovery from depression (τ_{rec}, 100ms). Normalized amplitudes. Frequency of stimulation for both synapses was 40Hz.

the most complex signals which could bind events over long time periods (Figure 5.6). An analogy is the binding of many letters to construct complete words, many words to construct complete sentences or many sounds to construct a complete sound percept. The fourth principle of the function of nonlinear synaptic transmission is therefore that the amount of temporal information packaged depends on the complexity of the interactions within the synaptic machinery and how rapidly the interactions equilibrate with respect to the input.

Fourth Principle of the Function of Nonlinear Synapses:

The amount of temporal information packaged depends on the rate that synaptic transmission equilibrates with respect to the input.

5.8   Classes of Temporal Information Packages

The particular temporal relationships between APs, which nonlinear synapses are so sensitive to, can be defined by examining synaptic transfer functions (input-output relationships) (Tsodyks and Markram, 1997; Markram et al., 1998a). Two basic transfer functions are considered; the steady state synaptic response as a function of AP frequency and the transient response at time t after a stimulus as a function of preceding AP frequency and increment in frequency. The tremendous potential significance of the steady state synaptic transfer function was recognized in a few of the earliest publications on synaptic transmission and some of the conclusions we now reach by examining this transfer function were in fact reached then (see Curtis and Eccles, 1960). Unfortunately, the subsequent trend to simplify synapses overlooked this approach.

The steady state synaptic response as a function of AP frequency is typically non-linear with frequency and the transient transfer function (tTF) is nonlinear with respect to the initial frequency and the increment in frequency. Unlike the steady state TF (stTF), the tTF cannot be simply described since a different function can be derived for a different time point or time window from the moment of a stimulus until steady state is reached; depending on the degree of digitization of synaptic parameters (Appendix I, Digitization of Synaptic Parameters) there could be an infinite number of different tTF for a given synapse. Since TI is packaged during the non-equilibrium period of synaptic transmission, the full complexity of the contents of these packets is represented by the tTF. The stTF is, however, the most practical since the frequency regimes in which specific classes of TI are primarily constructed, can be defined.

The steady state synaptic response of a typical depressing synapse between pyramidal neurons decreases progressively as the frequency (f) of stimulation increases (Figure 5.7).

Beyond a specific frequency (defined as the limiting frequency, λ), this decrease in synaptic response amplitude is approximately proportional to $1/f$ (Figure 5.8). This essentially means that the rate of transmission (product of synaptic response and frequency) saturates beyond a specific frequency. This effect was first observed in the late 1950s and most completely described by Curtis and Eccles (1960), by examining the steady state synaptic transfer function of motor neuron responses, and the specific frequency at which saturation occurred was referred to as the "critical frequency". Later theoretical studies also inferred the 1/f rule of synaptic depression (Appendix B, Functional Significance and Appendix J, Steady State).

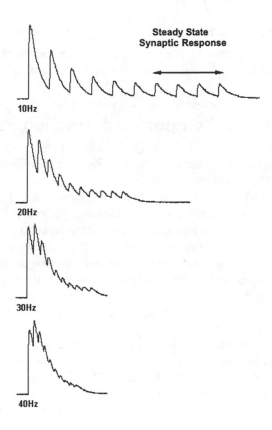

Figure 5.7. Classical response of a depressing synaptic connection to different frequencies of regular trains of APs.

The consequence for transmission is that when f changes, for example, from 10Hz to 40Hz, the average current injected into the target cell is the same after steady state is reached as it was before the change in f. When the frequency changes, however, due to a finite time period required for synaptic responses to depress to steady state level, the synaptic responses during the non-equilibrium period are higher than they will be at steady state while the frequency of stimulation is higher. For a *population* of synapses, the product of the summed synaptic responses and f therefore yields a transiently large net response starting at the moment the frequency changes which lasts for the duration of the non-equilibrium period of the synapse (Figure 5.9).

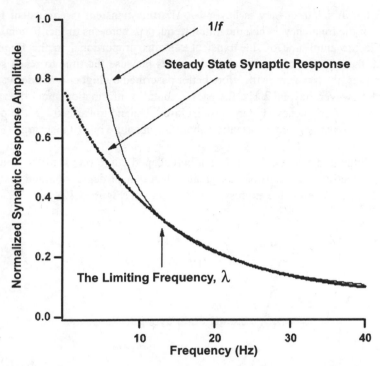

Figure 5.8. Steady state synaptic transfer function for a depressing synapse (see Tsodyks and Markram, 1997).

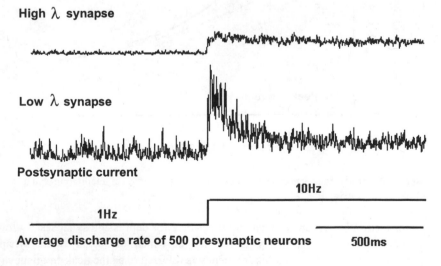

Figure 5.9. Transmitting rates and derivatives. Computed synaptic current generated by 500 presynaptic neurons via slowly (upper trace) depressing synapses with low average U and hence high λ values and fast (middle trace) depressing synapses with high average U and hence low λ values (see Tsodyks and Markram, 1997).

When the initial frequency is beyond λ Hz, this transient is equivalent to the derivative of the frequency. When the initial frequency increases further beyond λ Hz then, while the amplitude of the transient remains proportional to the frequency increment, the duration of the transient decreases because the time to reach steady state decreases; the transient is therefore better described by higher order derivatives. In general, however, beyond λ Hz, the packet contains information about the rate of change of the AP frequency (Figure 5.9). Below λ Hz and towards 0 Hz, the packet contains progressively more information about the absolute rate of discharge (Figure 5.9).

For facilitating synapses, the stTF is a bell-shaped curve provided that the time constant for facilitation is sufficiently greater than that of depression and that U is sufficiently small to allow the expression of facilitation (Figure 5.10).

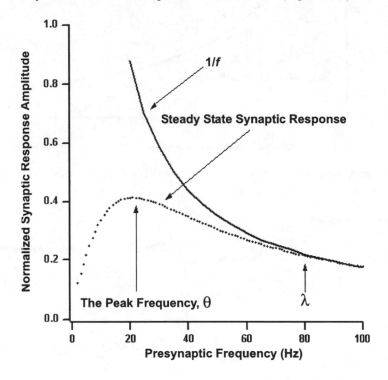

Figure 5.10. Steady state synaptic transfer function for a typical facilitating synapse (see Markram et al., 1998a).

λ for facilitating synapses is much higher than for depressing synapses. At lower frequencies, a special frequency is reached at which the steady state synaptic response is at a maximum and this frequency is referred to as the peak frequency (θ). Curtis and Eccles (1960), referred to this frequency as the "optimal frequency". The peak frequency is characteristic of the synaptic parameters (Appendix J, Steady State). For frequencies around θ Hz the synapse transmits almost linearly as a function of frequency (f * response amplitude) and for frequencies below θ Hz and towards 0 Hz the postsynaptic response reflects the product of the discharge rate and the integral of the discharge. θ and λ therefore delineate three basic frequency regimes in which elementary features of the AP train are packaged. The three main elementary packages of TI created are AP discharge rates, derivative of rates and integrals of rates. For Poisson spikes, the synaptic response may contain any combination of elementary features potentially allowing unique multidimensional representations of network activity by the precise sequences of APs.

5.9   Emergence of the Population Signal

While the packet of TI that is constructed by nonlinear synapses can be derived analytically or in simulations, if the postsynaptic voltage response to a single train of APs is examined, then the packet is not evident. For example, at a depressing synapse the synaptic response only decreases when the frequency increases further (Figure 5.11). How then can the synapse contribute to the discharge of the neuron at the moment of an increase in frequency?

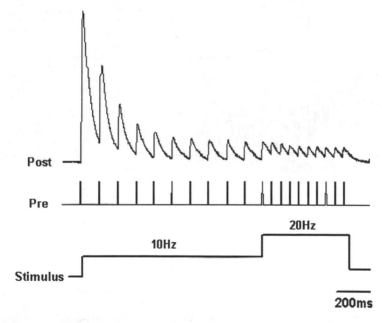

Figure 5.11. Synaptic response to a frequency transition of a regular train of APs.

For depressing synapses, the derivative of the frequency is contributed to the population signal because the synapse is driven at a higher frequency before the synaptic response has depressed, but the derivative of the instantaneous frequency (1/ISI) of an irregular spike train is noisy and hence the packet of information is buried within the synapse's own noise. An average of several similar inputs from other synapses is required to average out the noise and extract the signal. The TI transmitted by a single synapse is therefore entirely dependent on associative synaptic input from *other neurons* in order to be realized and hence the term "population signal". The more presynaptic neurons involved in the synaptic input the clearer these population signals are and the more precisely the discharge activity correlates with specific spatio-temporal patterns of the network (Figure 5.12).

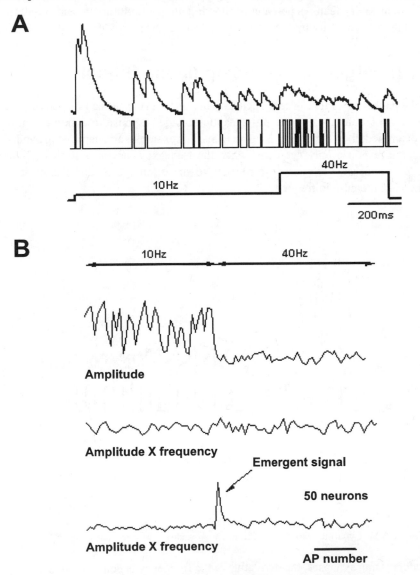

Figure 5.12. Emergence of the population signal. (A) Synaptic response to a frequency transition of an irregular train of APs. (B) (Upper trace) The amplitude of the synaptic response evoked by each AP. (Middle trace) The upper trace multiplied by the average frequency of stimulation (10Hz and 40Hz) for a single neuron and (Lower trace) for 50 neurons (see Tsodyks and Markram, 1997).

Typical amplitudes of recovered or rested synaptic responses between pyramidal neurons is around 1mV, indicating that about 10 neurons could bring another neuron to threshold and about 50 would strongly drive the neuron. This is a configurable variable since at high frequencies the synaptic responses can decrease to below 100μV and then up to 500 neurons may be required to produce an equivalent discharge. Shunting inhibition is one potential mechanism that could ensure pop out of the population signal by down scaling the strength of each input and allowing many neurons to drive the cell. Shunting inhibition could therefore serve to dynamically tune how precisely neurons will respond to the spatio-temporal patterns of activity within the network. Regarding inhibition as a game of balancing excitation and inhibition is therefore probably a gross oversimplification (Appendix K, Inhibitory Synapses).

Temporal information about the non steady state AP frequencies can therefore be generated by single synapses, but this information cannot be expressed via the voltage response in terms of a contribution to discharge the neuron unless the input is appropriately timed to be compatible with associative synaptic input from the presynaptic population. If probabilistic synaptic transmission is also considered, then the packet of TI is buried even deeper in the synapse's own noise and this function of stochasticity could be formalized as an estimation or prediction of the mean packet of TI for a single input as well as for all associative inputs.

Fifth Principle of the Function of Nonlinear Synapses:

Temporal information encoded in nonlinear synapses requires associative input from the presynaptic population in order to be extracted.

In order for a packet of TI to get through to the postsynaptic neuron and contribute to the activity of the neuron, it must therefore be relevant to the ongoing activity within the network. The sixth principle of the function of nonlinear synaptic transmission is therefore that temporal information encoded by synapses predicts the relative timing of AP discharge across the presynaptic population of neurons. Since neurons are interconnected recurrently, this principle can be extended to where single synapses predict y-network state from x-presynaptic AP discharge pattern. The particular feature of the network activity that is predicted by the synapse depends on the values of the synaptic parameters; in other words, the way in which the synapse is tuned determines what aspect of the network activity is predicted.

Sixth Principle of the Function of Nonlinear Synapses:

Temporal information encoded by synapses predicts the relative timing of AP discharge across the presynaptic population of neurons and a particular network state in a recurrent network.

5.10   Recurrent Neural Networks

As nothing can exist if it does not combine all the conditions which render its existence possible, the different parts of each being must be co-ordinated in

such a manner so as to render the total being possible…this is the principle of
the conditions of existence, vulgarly called the principle of final causes.

George Cuvier, 1817

The emergence of animate systems is subserved by a progressive increase in the
interdependency of functions of the elementary components of a system; all that distinguishes animate from inanimate involves an obvious collaboration of functions.
Biological systems are therefore massively recurrent at the genetic, molecular, cellular, extracellular and network levels. One of the most powerful anatomical designs
that render the activity of each neuron dependent on all other neurons in some way is
recurrent microcircuitry. In fact, when one considers simply that APs in the neocortex
occur as a function of other APs, then the usual rate versus the time coding argument
pales, since multidimensional and multiple time-scaled correlations must exist
between the APs emitted by one neuron and those emitted by other neurons. Anatom-
ically, each neuron is also related to each other neuron in a multilevel manner with
various ratios of 1st, 2nd, Nth order synaptic connectivity. Due to this recurrent
microcircuitry, no neuron is more than a few synaptic junctions away from any other
neuron in the nervous system (Appendix L, Lack of Boundaries). Recurrent microcircuitry is genetically prescribed and is probably modified considerably during learning. Recurrent microcircuitry can therefore be seen as containing all the information
acquired by the particular animal species during evolution and during the personal
history of the animal that pertains to the field of specialization of the circuit. The
question is how is this information retrieved and how does it subserve, influence or
determine normal perception?

5.11   Combining Temporal Information in
Recurrent Networks

Stimuli projected through a feed-forward pathway into a recurrent network initially
drives the neurons according to the topography of the input and not according to the
recurrent architecture. In other words, at the very onset of a feedforward input the
relationship between the APs discharged by each neuron contains some information
about the stimulus, but little about the structure of the recurrent microcircuitry. With
the progression of time, however, the discharges reflect progressively more information about the recurrence as the interactions between the neurons begin to exert their
effect. This emergent recurrent information (RI) rises to a maximum level as the
recurrent microcircuitry is engaged. RI is then progressively inserted into AP trains
from the onset of each new stimulus (Appendix M, Speed of RI Accumulation). Full
expression of RI requires interaction between the neurons of the network and hence
involves sequences of APs. Since synaptic dynamics learn and align to these
sequences of inputs one can infer that synaptic dynamics are tuned to predict RI
rather than any specific stimulus. The seventh principle of the function of nonlinear
synaptic transmission is therefore that nonlinear synaptic transmission predicts the

information embedded in the recurrent architecture that has been acquired during evolution and during the personal history of the organism.

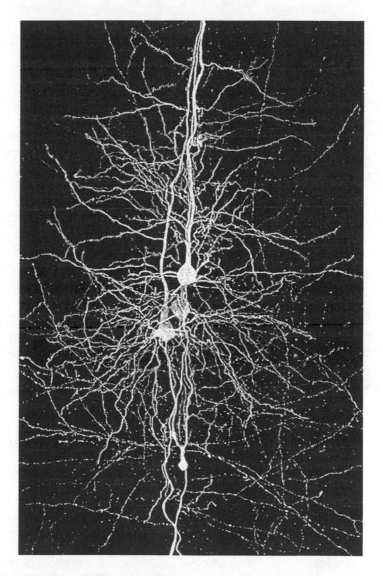

Figure 5.13. Three layer 5 pyramidal cells in rat somatosensory cortex. They are connected to each other representing the recurrent network in the neocortex. A light microscopic psuedo-color image of the somatic region was taken from three biocytin-filled pyramidal cells. Image processed by Yun Wang. © Yun Wang, 1997.

It is hypothesized that regardless of the stimulus, synapses are optimally tuned to predict only those AP patterns that would contain maximal RI. Nonlinear synapses could therefore mature to predict all past knowledge embedded in the network.

Seventh Principle of the Function of Nonlinear Synapses:

Nonlinear synaptic transmission predicts information embedded in the recurrent architecture.

5.12   Organization of Synaptic Parameters

One of the most striking and elegant features of the neocortical microcircuitry is the incredibly precise organization of synaptic dynamics wherein each neuron maps onto specific target neurons using specific forms of synaptic dynamics. The dynamics of synapses can be classified broadly into depressing (D-type) and facilitating (F-type) synapses. A more detailed classification considers the ratio of facilitation and depression time constants. In this scheme, F1 type synapses are synapses with much greater time constants of facilitation than depression (typical facilitatory synapses). The F2 type synapse is essentially the opposite with long time constants of depression and short time constants of facilitation and the F3 type synapse has similar time constants for facilitation and depression. The synapses between pyramidal neurons are depressing synapses with essentially no facilitatory component (D-type) in the young neocortex (pure depressing synapses) and a fast time constant of facilitation in the mature neocortex (glutamate-F2-type). Synapses from pyramidal cells onto interneurons, between interneurons and from interneurons onto pyramidal neurons can be F1, F2- or F3-type synapses. More than 30 different anatomical-electrophysiological types of interneurons have been defined and each of these receives and forms specific types of synapses onto different targets revealing an intricate functional map of synaptic dynamics. This map follows broad principles:

Principle of mapping glutamatergic synaptic dynamics:

1.   The same axon can form different types of synapses.

2.   The same neurons can receive different types of synapses.

3.   Precise mapping of the type of synaptic dynamics from pyramidal neurons onto other target neurons.

4.   Mapping of glutamatergic synaptic dynamics according to pre- and postsynaptic anatomical and electrophysiological properties.

5.   Heterogeneity of synaptic dynamics within a class of synaptic dynamics onto the same type of target neuron.

Principle of mapping GABAergic synaptic dynamics:

1.   The same axon can form different types of synapses.

2. The same neurons can receive different types of synapses.

3. Precise mapping of the type of synaptic dynamics from interneu-
    rons onto other target neurons.

4. Mapping of GABAergic synaptic dynamics according to pre- and
    postsynaptic anatomical and electrophysiological properties.

5. Homogeneity of synaptic dynamics within a class of synaptic
    dynamics onto the same type of target neuron.

5.13   Learning Dynamics, Learning to Predict

The evidence for learning of synaptic dynamics is provided by the large dynamic
ranges of values of synaptic parameters. For glutamatergic synapses, triple and quad-
ruple recordings revealed a large degree of heterogeneity in multiple synaptic parameters of synapses that are established by a single axon of a neuron onto different
target neurons. Heterogeneity in all four synaptic parameters (A, U, τ_{rec} and τ_{facil})
was found indicating that each synaptic connection established onto a target neuron
has potentially unique transfer functions. Synapses formed onto the same type of target neurons, fell within the same class of synapse (e.g. facilitation or depression)
indicating that a common denominator of transfer functions exist for inputs from the
same class of presynaptic neurons. Recent studies revealed that different classes of
synapses can be established on the same pyramidal neuron or the same interneuron,
which provides for maximal combinatorial complexity in organizing nonlinear synapses between different pre- and postsynaptic neurons in a network; in other words
there potentially exists as many types of synapses as there are classes of pre- and
postsynaptic neurons.

For excitatory, glutamatergic synapses, considerable heterogeneity in the values
of the synaptic parameters was also found within a synaptic class indicating that the
synaptic transfer to each target neuron is still unique. This contrasts sharply with the
organizing principle found for inhibitory synapses where almost perfect homogeneity
of U, τ_{rec} and τ_{facil} were found when the axon diverged to contact many neurons of
the same class. For excitatory synapses these results suggest two general principles
that underlie the generation of synaptic dynamics: (a) the class of synapse is set in an
activity-independent (innate, genetic setting) manner depending on the genetic nature
of the pre- and postsynaptic neurons while (b) the history of activity between the pre-
and postsynaptic neuron sets the precise values of the synaptic parameters in an *activ-
ity-dependent* manner (Figure 5.14).

These data indicate that multiple synaptic parameters of single synapses are modifiable providing the biological foundation for intricate tuning of synaptic predictions
that may be required for each synapse to predict the specific network activity pattern

from its particular location on a neuron and in the network that is orchestrated with
the predictions of all other synapses.

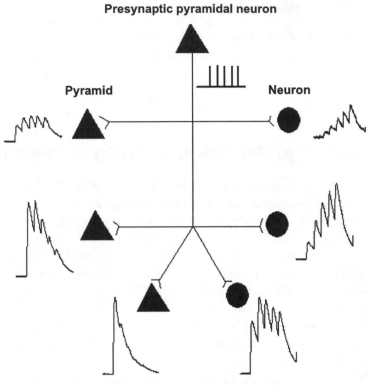

Figure 5.14. Differential synaptic transmission via the same axon. The rules that are obeyed
are: depressing synapses onto pyramidal neurons and facilitating synapses onto certain classes
of interneurons; heterogeneity of synaptic dynamics within each class (see Markram et al.,
1998a).

5.14   Redistribution of Synaptic Efficacy

Based on early theories of learning mechanisms, experiments examining plasticity of
synaptic transmission have interpreted all changes in transmission as a change in the
strength of transmission. Linear synapses do not compute and are subject to only
potentiation and depression (gain change), but nonlinear synapses do compute and
are therefore subject to a change in the gain as well as the contents of the computation – i.e. a change in the form of temporal integration operation. Changing any of the
kinetic parameters causes differential changes in the transmission of each AP in a
train and therefore alters the contents of the information transmitted by the train; this
phenomenon is referred to as "redistribution of synaptic efficacy" (Figure 5.15).

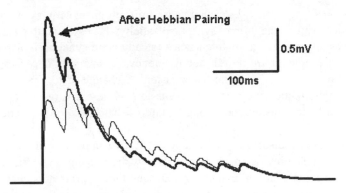

Figure 5.15. Redistribution of synaptic efficacy. The classical phenomenon that is observed between neocortical pyramidal neurons when synaptic responses to regular AP trains are examined before and after Hebbian pairing. Redistribution in this case is caused by an increase in U (see Markram and Tsodyks, 1996).

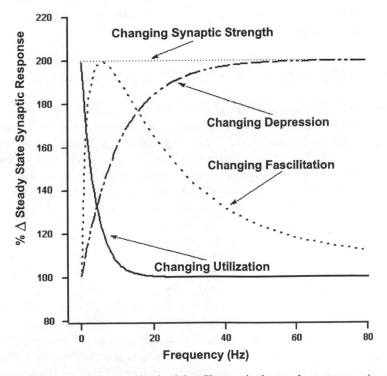

Figure 5.16. Four classes of synaptic plasticity. Changes in the steady state synaptic response for different frequencies when either A, U, τ_{rec}, or τ_{facil} are changed. This illustrates that changing A changes synaptic transmission uniformly for all frequencies, changing U changes only low frequency synaptic transmission, changing τ_{rec} only changes high frequency synaptic transmission and changing τ_{facil} changes transmission within a band of frequencies. Experimental evidence suggest that all changes are possible at the same synapse. Y-scale normalized to 200% maximum change (see Markram et al., 1998b).

Changing U, τ_{rec} and τ_{facil} cannot cause uniform changes in synaptic transmission. In order to study the form of synaptic plasticity, it is essential to consider synaptic transmission for a range of frequencies (steady state synaptic transfer function) and not just a single arbitrarily selected frequency. A change in U causes a selective change in low frequency synaptic transmission. A change in τ_{rec} causes a selective change in high frequency synaptic transmission while a change in τ_{facil} causes a selective change in transmission during an intermediate band of frequencies (Figure 5.16).

These forms of redistribution of synaptic efficacy can be quantified as changes in λ and θ, which also provide an assessment of how tTFs change. The first principle of synaptic learning is therefore that changes in the kinetic parameters change the contents of the temporal information transmitted while a change in absolute synaptic strength changes its gain.

First Principle of Synaptic Learning:

Changes in parameters that contribute to the dynamics of synaptic transmission, change the contents of the information transmitted while a change in absolute synaptic strength changes the gain.

5.15   Optimizing Synaptic Prediction

A comprehensive algorithm to change the probability of neurotransmitter release (or U) by considering the precise relative timing of APs in the pre- and postsynaptic neurons was established (Senn et al., 1997). This algorithm is based on experiments which showed that the AP propagating back into dendrites triggers an increase in p when the AP collides with incoming synaptic input (i.e. if the synaptic input contributes to the AP) while the backpropagating AP triggers a decrease in p when the synaptic input arrives in the dendrite a few milliseconds after the AP. The principle embodied in this algorithm is a causal/acausal reward/punishment principle between the input and output (James' and Hebb's principles) (Figure 5.17).

It is important to note that if this algorithm is implemented explicitly (i.e. the relative AP timing is directly translated into a value to increment the value of a synaptic parameter), then synaptic convergence and stability of single synapses is not possible without other constraints (such as network stabilizing effects). This is a trivial consequence of the synapse being driven to the value where it is the most unstable because a small AP timing error will lead to the largest change. For example, when synapses are strengthened it also improves its positive temporal correlation with the output of the target and gets even stronger – reaching the near zero AP time difference where a tiny error will cause a very large change. Interestingly, the stable point for a network with such a learning algorithm is where all activity is maximally desynchronized (i.e. in this implementation, a temporal learning rule would generate a rate favored code!). Aside from network solutions to this problem, the problem can be solved by implementing the learning rule via second messengers in order to generate higher order dynamics for the AP timing to interact (Senn et al., 1997).

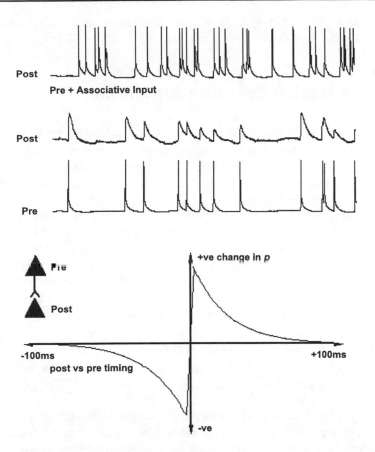

Figure 5.17. A fine line between Hebbian and anti-Hebbian synaptic plasticity. The function depicted in the lower graph allows the prediction of the evolution of p as a function of relative timing of pre- and postsynaptic AP trains (see Markram et al., 1997b).

The effect of applying this algorithm through a "biophysical buffer" is that simultaneously active synapses contribute to the discharge of the neuron and p increases to a level that corresponds to the *degree* of correlation between the input from a particular presynaptic neuron and the output of the target neuron. This reduces λ and lowers the frequency at which the synapse will transmit the derivatives. The consequence is that the synapse must now predict more accurately (with fewer APs and at lower frequencies) the precise time that the presynaptic population is changing frequencies. If the prediction is inaccurate then p is reduced, λ shifts to the right and the synapse slides backward into a mode where only average rates are predicted. Regardless of the direction of the change, synapses adjust their prediction to what is possible given the relationship of the presynaptic neuron with the target neuron in the context of the entire network; in the algorithm p converges to a value characteristic of the degree of correlation between the discharge of the pre- and postsynaptic neurons. Decreasing or increasing p (also τ_{rec} and τ_{facil}) is therefore not synonymous with changing the gain of transmission, but rather with altering the contents of the temporal information transmitted to optimize the synaptic prediction.

5.16   A Nested Learning Algorithm

Changing any one of the kinetic parameters (U, τ_{rec}, τ_{facil}) has an impact on transmission that depends on the values of the other synaptic parameters. The ultimate algorithm for learning synaptic dynamics should therefore regulate each synaptic parameter as a function of the other parameters and a supra-algorithm is therefore predicted in which these interrelated algorithms are nested. The primary effect of such an algorithm would be to establish specific correlations between synaptic parameters and, because the algorithm is driven by success and failure of synaptic predictions, this also means that each synaptic parameter in each synapse will be correlated with each synaptic parameter in each other synapse in the network. The second potential principle of synaptic learning is therefore that synaptic learning of nonlinear synapses serves to correlate the values of synaptic parameters within a single synapse and across all synapses in the entire network.

Second Principle of Synaptic Learning:

Nonlinear synaptic transmission must be aligned across all synapses throughout a network.

A multidimensional vector of values of synaptic parameters is therefore proposed to evolve with repeated activation of the network by various stimuli towards an optimally mature form, which will allow each synapse in the network to predict a compatible state of the network under virtually any conditions of stimulation. If this network state can be reached, then RI would be maximal and this condition would be a state of maximal synaptic synergy or orchestration (Appendix N, Network Efficiency). An optimally mature form of the multidimensional vector is therefore achieved when the network has been tested sufficiently by diverse stimuli to allow synapses to reach the optimal synaptic predictions of the knowledge embedded in the network.

In this theory, the primary objective of synaptic learning is not to improve the analysis of any particular stimulus, but to allow the synaptic predictions to embody all the information *already* embedded in the recurrent microcircuitry; i.e. the network learns about itself. Once this has been achieved, any stimulus that succeeds in evoking successful synaptic predictions will automatically be compared to the pre-existing knowledge and be defined; potentially even stimuli that were not used in the training period to mature the synaptic predictions. Alien (entirely novel) stimuli would probably not evoke any response whatsoever and can not be perceived by the network, and the nervous system must therefore learn in small increments of novelty. An even stronger claim could be made that a neural network cannot even learn to perceive stimuli that are truly foreign since the acquisition of new knowledge requires the recruitment from a vast genetic reservoir of information and therefore whatever can be learned to perceive is already known to some degree at the genetic level. The basic organization of the nervous system of any particular species of animal is therefore likely to impose unassailable limitations in perception.

The strength of synaptic connections is an integral part of the "structure of recurrence" and plasticity of synaptic strength therefore provides a means by which to acquire new knowledge (as has been proposed and implemented in learning theories for decades). Different synaptic parameters, however, serve different functions; the synaptic strength could be viewed as delineating the potential activity space that can be explored by the dynamics (the borders or framework of activity patterns) while the dynamics could be seen as describing the potential routes that can be taken to explore this activity space. The activity space for even a single cortical column is incredibly large; only considering each neuron as a single degree of freedom means these columns are processing information in at least a 10,000 dimensional space. A prediction from this theory is that changes in structure of the network must be accompanied by iterative changes in the dynamics of the synapses to enable the network to explore this complex space.

Third Principle of Synaptic Learning:

Nonlinear synapses mature to predict network structure and generate maximally synergistic responses to AP patterns that would result from maximal interactions between neurons in a recurrent network.

Fourth Principle of Synaptic Learning:

Synaptic strength defines the borders of the possible activity space of a network while synaptic dynamics determine the possible trajectories of multidimensional activity patterns that are possible within these borders.

5.17   Retrieving Memories from Nonlinear Synapses

The principles proposed above may help to understand the physical constraints for storing and retrieving information in synapses with complex synaptic dynamics. Traditionally, only plasticity of synaptic strength has been considered, but now we know that all synaptic parameters (strength, p, depression and facilitation) are plastic parameters. The requirements for storing and retrieving a memory in the strength of a synapses are straightforward; change the gain and each subsequent AP will generate a response that will reflect this change and thereby add this to the network activity. The requirements for storing and retrieving information in the other "kinetic parameters" is much more complex. For example, if a memory is stored by adjusting the degree of synaptic depression or facilitation, then only a train of APs will be able to retrieve this memory since the depression and facilitation must be engaged. Most have thought it is also straightforward for p, but when p is changed, so is the effective engagement of depression and facilitation, thus the profiles of responses to sequences of APs can be radically altered. The consequences of storing information in the value of p, is therefore also extremely complex.

Fifth Principle of Synaptic Learning:

Retrieving memories from nonlinear synapses requires trains of APs where the number, frequency and structure of the train depends on the values of the synaptic parameters.

Consider the network case where memories are stored in the values of all four of these parameters. When the network is driven in a feed-forward manner, the first spikes caused by the stimulus will generate responses that could come from virtually any settings of these four parameters. Only when each connection transmits the required number of APs will the responses become characteristic and unambiguous for each synapse; only then will the response generated in the network reflect the synaptic settings. Retrieving memories from nonlinear synapses in the classical model of memory where the activity patterns must be recreated is therefore a highly non-trivial task.

An alternative solution to this serious problem is an entirely different view of information processing, learning and memory where the goal is not to recreate any particular activity pattern, but merely to maintain interactions between neurons along specific trajectories of activities. Synaptic settings prescribe trajectories of activity patterns and at the onset of a stimulus, a large number of possible trajectories can be "chosen" by the network. The secret to memory retrieval may then lie in the hands of ingenious readout mechanisms that are able to read the *trajectories* of activity patterns from the current state of the network rather than a network relying on any *specific* and *stable* activity pattern.

5.18   Conclusion

The chapter proposes elementary principles for information processing, learning and memory in recurrent neural networks with nonlinear synapses. This theory of the function of synapses emerged and is still emerging as these principles are considered further. The most useful aspect of this theory is not as an explanation per se, but as a source of predictions that challenge our understanding of the nervous system and challenge models of the nervous system to incorporate and understand the full complexity of the microcircuit.

The emergent theory is that recurrent microcircuits are essentially continual temporal processing devices. The multiple interacting processes and time constants of interactions that determine the output of synapses render nonlinear synapses as ideal temporal processing devices. Temporal features of environmental input can therefore be encoded at the synaptic level. Nonlinear synapses are extremely diverse with specific types of nonlinearities between specific types of neurons, indicating that a connection between two specific types of neurons is devoted to a specific form, feature and duration of temporal integration. These temporal feature maps appear to be largely organized genetically. Plasticity of the dynamics of synaptic transmission allows fine-tuning of the specific temporal processing between two neurons.

This analysis shows that neurons receive temporal information which is precisely mapped onto the dendritic tree of neurons allowing each neuron to generate an extremely high dimensional space-time representation of the input it receives. Each component of a neuron that dynamically exerts an influence on the neuron's activity (ion channels, receptors, second messengers, etc.) is a potential degree of freedom (dimension) and hence neurons process information in an essentially infinitely high dimensional space. This high dimensional representation is collapsed in the output spike train into a single dimension – time.

This temporal information is re-represented in a spatial map in the recurrent network by the axon diverging and contacting hundreds or thousands of neurons in specific locations in the network, and on specific locations on each neuron's dendritic tree. The iteration starts again, only now the information holds even more information about each of the neuron's processing and the connectivity structure. With each step of recurrent processing, therefore, information that is embedded at all levels of the microcircuit (molecular to circuit level) is retrieved and inserted into the AP train. The essential principle that is proposed, therefore, is that recurrent microcircuits process temporal features (time is created in recurrence) and this information is represented and held continuously in spatial maps of activities of single neurons and the network. Spatial information about stimuli and the world, on the other hand, is superimposed and processed within topographic macroscopic and microscopic maps of projection pathways.

Finally, this theory predicts that recurrent microcircuits generate a continuous neural environment (i.e. no digitization in the neural representation). This environment is a "raw" or "unbiased" representation of the stimulus environment and can therefore not be described with any one code (Appendix O, The Binding Problem of the Binding Problem). The information is represented across an incredibly large number of dimensions that seems crucial to allow an always changing activity pattern to realize a stable perception and perfect memory recall. While readout mechanisms, such as a single axon or a projection pathway, collapse these dimensions to generate a specific output that can be described with a specific neural code (such as rate, time, synchrony, synfire etc.) the actual neural representation within recurrence is an unbiased multidimensional representation of all past experience of the organism.

5.19   Acknowledgements

I would like to acknowledge the great collaboration with Misha Tsodyks who has contributed significantly to many of the derivations made in this chapter. I would like to thank the foundations that supported the construction of the laboratory and the subsequent experimental and theoretical work described in this chapter: The Henry and Ann Reich Research Foundation; The Dr. Perl H. Levine Foundation; The Levinson Family Foundation; The Abramson Family Foundation; The Grodetsky Center for Higher Brain Function; The Israel Academy of Sciences; The United States Office of Naval Research; The Minerva Foundation; The United States-Israel Binational Science Foundation; The French-Israel Binational Science Foundation; The Human Frontiers of Science Program. HM has been the incumbent of the Joseph D. Shane Career development chair.

Appendix A: Sherrington's Leap

Sherrington made the quantum leap from an electrical synapse to a chemical synapse and introduced the term "synapse" (Sherrington, 1897).

Appendix B: Functional Significance

Several groups have considered the functional significance of nonlinear synaptic transmission. Nonlinear synaptic transmission has been observed in some of the earliest recordings of synaptic potentials (Feng, 1940; Eccles et al., 1941; Eccles, 1946; Lloyd, 1949; Liley and North, 1953; Curtis and Eccles, 1960; Dudel, 1965; Mallart and Martin, 1967). In particular, Eccles seemed to realize the potential significance of examining transmission using trains of APs and that a synapse can only be completely characterized by testing its response at multiple frequencies. In 1960, Eccles specifically pointed out two features of synapses that were recently "rediscovered" in neocortical synapses (Tsodyks and Markram, 1997; Markram et al., 1998a; Abbott et al., 1997) and that are described in this chapter; a "critical frequency" where neurotransmitter release is inversely proportional to the frequency of stimulation and an "optimal frequency" at which the synaptic response is the maximum (Curtis and Eccles, 1960). Later studies on the neuromuscular junction also pointed out specifically the potential importance of frequency dependent synaptic transmission (Magleby, 1987).

The potential of a single neuron, a command neuron (Wiersma and Ikeda, 1964), with differential nonlinear synaptic transmission at each output, to enable differential control of muscle groups was also realized early on (Parnas, 1972; Grossman et al., 1973) and developed further to explain how single neurons can control complex behavior (Atwood and Wojtowicz, 1986; Frost and Katz, 1996; Dityatev and Clamann, 1998). Synaptic depression has also been central to studies on habituation (Byrne, 1982) and rapidly changing synapses formed the basis of a form of associative principle of knowledge representation (von der Malsburg and Bienenstock, 1986).

Nonlinear synaptic transmission has been extensively incorporated as "gates" into models by Grossberg since the 1960s – much too extensive to be reviewed here (see Grossberg, 1980, 1984a, 1992). Briefly, according to Grossberg, the law of depression (Grossberg also applied the 1/f law by assuming that the release rate saturates; see Grossberg, 1969a) is the simplest law for "unbiased transduction by a chemical transmitter in which the transmitter recovery rate is slower than the input fluctuation rate" and Grossberg therefore proposes that rates of recovery of transmitter are "chosen" relative to the input fluctuation rate by evolutionary selection and makes the case by showing that specific relative rates are fundamental to many functions of recurrent and non-recurrent networks (personal communication). Grossberg implemented nonlinear synaptic transmission to explain feature binding (Francis et al., 1994), associative learning in recurrent and non-recurrent networks (Grossberg, 1969b), reinforcement learning (Grossberg, 1972), adaptively timed reinforcement learning (Grossberg and Merrill, 1992), memory (Grossberg, 1976), motion perception (Francis and Grossberg, 1996), in building the circadian clock (Carpenter and Grossberg, 1983) and in mental disorders (Grossberg, 1984a), to name but a few. In terms of synaptic learning, Grossberg's models for associative learning constitute a law for "unbiased spatial pattern learning in recurrent and non-recurrent networks using chemical transmitters whose processes fluctuate at a finite rate", essentially meaning that nonlinear synaptic transmission allows for an automatic normalization of AP patterns to allow the "unbiased" association of patterns.

Taken together, the Grossberg studies clearly show that synaptic depression (and facilitation in some cases) is an extremely powerful design principle in building models of nervous function. This testifies to tremendous insight by Grossberg since most of these derivations where achieved without much quantitative data of synaptic properties. With the emergence of quantitative synaptic properties, however, it is now important to incorporate some fundamental findings (rules of organization of nonlinear synapses between specific neurons, amplitudes and ranges of time courses typically found; different classes of nonlinear synapses) in order to achieve a comprehensive understanding of why the nervous system implements nonlinear synaptic transmission.

More recently, Markram and Tsodyks (1996) showed that synaptic modifications can involve a change in the contents of the information transmitted by synapses rather than merely the gain of the transmission. This initiated a series of studies aimed at understanding what "contents" of information could actually mean, one showed the 1/f law for neocortical synapses and derived two elementary features of a presynaptic AP train that can be transmitted by synapses (derivatives and rates) (Tsodyks and Markram, 1996, 1997) and another reported the peak frequency for facilitating synapses and derived the third elementary feature, the integral of discharge rates (Markram et al., 1998a). At about the same time Liaw and Berger explored the importance of "dynamic synapses" in isolating dynamic signals (voice recognition) within noise (Liaw and Berger, 1996). Abbott and colleagues also independently found the 1/f law for neocortical synapses and further showed that this underlies an "automatic gain control" mechanism that allows slow and fast firing afferents to produce equivalent responses, and that synaptic depression increases the sensitivity of neurons to the precise pattern of APs in the afferents (Abbott et al., 1997).

Appendix C: Visual Patch Recordings

In order to study synaptic connections between pre-specified neurons in the neocortex repeatedly it is necessary to visually locate these neurons prior to obtaining recordings. This approach is enabled by the recent application of video-enhanced infrared differential interference contrast microscopy (IR-DIC) in brain slices to visually preselect neurons for recordings (Markram and Tsodyks, 1996; Markram et al., 1997a). The advantages of this approach are that whole-cell recordings can be obtained from both pre- and postsynaptic neurons enabling precise control over the AP activity patterns in the coupled neurons. Subsequently the detailed microanatomy and putative number and location of synaptic contacts can be determined.

Appendix D: Biophysical Basis of Parameters

The parameter A:
  A is the maximum response that can be generated by a synaptic connection; i.e. when $p=1$ at all synapses forming the connection. A is therefore equivalent to the quantal size multiplied by the number of release sites and an electrotonic attenuation factor. This measure of the absolute synaptic strength reflects the potential of the synaptic input to directly influence the discharge of an AP (see also Markram et al., 1998b).

The parameter U:
  U is equivalent to p provided it can be established that the mechanism of frequency dependence is located purely presynaptically. If postsynaptic receptor desensitization contributes to frequency dependence, the U reflects both p and the effect of receptor desensitization. U is therefore a functional parameter that does not depend on assumption of the mechanism of depression.

The parameter τ_{rec}:
  τ_{rec} can be determined directly from the experimental traces and used to constrain the model fitting procedure. The biophysical correlate of recovery from depression could be vesicle depletion (Liley and North, 1953; Zucker, 1989) or recovery from a functional refractory period of release due to, for example, sequential decrease in AP-evoked Ca^{2+} influx (Zucker, 1989) or desensitization of the Ca^{2+}-induced release machinery. Perhaps only very slow rates of recovery from depression induced by intense activity are due to vesicle depletion (see Brodin et al., 1997), while a functional refractory period determines fast recovery from depression. Typically two rates

can be distinguished at neocortical synapses; a fast rate 100–2000ms and a slow rate of tens of seconds. We consider only the fast rate.

The parameter τ_{facil}:
The biophysical basis of facilitation has been studied extensively (Magleby and Zengel, 1982; Zucker, 1989; Kamiya and Zucker, 1994), mostly pointing to residual Ca^{2+} in the terminal as a general mechanism. The specific mechanisms underlying the various time constants for facilitation, however, are less clear (see Magleby and Zengel, 1982). The parameter, τ_{facil} could represent an average over the first two faster time constants of facilitation and perhaps part of augmentation described for data from the neuromuscular junction (Magleby and Zengel, 1982).

Appendix E: Single Connection, Many Synapses

Most connections between neurons in the neocortex are established by more than one synapse. For pyramidal to pyramidal connections this is typically between 4 and 8 (Markram et al., 1997b) and for connections from interneurons onto pyramidal cells between 10 and 30 (Y. Wang and H. Markram, unpublished; see Somogyi et al., 1998). The recording therefore provides an average of all the synapses and hence synaptic parameters reflect this average. This approach misses the potential computational complexity that may be involved in local regions of the neuron.

Appendix F: The Model

The phenomenological model (Tsodyks and Markram, 1997; Markram et al., 1998c) is based on two biophysical properties of frequency dependence. First, when the neurotransmitter is released, the site from which it was released cannot release again for some time (see Betz, 1970; Curtis and Eccles, 1960). The major component of the recovery of this synaptic refractory period appears to be a single stochastic process and can therefore be modeled as a single exponential. Second, no matter whether the neurotransmitter is released or not, each AP that arrives in the terminal sensitizes or facilitates the release by subsequent APs. The recovery from facilitation can be modeled approximately as a pulse-like increase in the probability of neurotransmitter release followed by a single exponential stochastic recovery (see Mallart and Martin, 1967; Magleby and Zengel, 1982; Zucker, 1989). The basic components of the model are:

1. The connection has an absolute synaptic efficacy (A), defined as the maximum response if all active sites released a transmitter simultaneously (i.e. when Pr is 1 at all release sites).

2.   The presynaptic AP utilizes some fraction of the A, defined as the Utilization of Synaptic Efficacy parameter (U).

3. U changes for each AP. The running value of U is referred to as u.

4.   Each AP causes a pulsed transient increase in U by an amplitude of $(U(1-u))$.

5. u decays with a time constant of τ_{facil}.

6.   The response is equivalent to the Effective Synaptic Efficacy (E) which inactivates virtually instantly (τ_{inac}).

7.   The Inactive Synaptic Efficacy (I) recovers with a time constant of τ_{rec} to add to the pool of Recovered Synaptic Efficacy (R).

The arrival of an AP utilizes a fraction (U) of the synaptic efficacy which becomes instantaneously unavailable for subsequent use and recovers with a time constant of τ_{rec}. The fraction of available synaptic efficacy is termed R. A facilitating mechanism is included in the model as a pulsed increase in U by each AP. The running value of U is referred to as u while U remains a parameter that applies to the first AP in a train between APs. u decays with a single exponential, τ_{facil}, to its resting value U.

From a resting state of the synapse, all the synaptic efficacy is available and the fraction that remains immediately after the first AP in a train is,

$$R_1 = 1 - U. \tag{A1}$$

During the AP train, each presynaptic AP utilizes further fractions of R at the time of its arrival. R therefore constantly changes because of subsequent utilization by APs, recovery of the unavailable synaptic efficacy with a time constant of τ_{rec}, and the pulsed increase in u caused by each AP. R for consecutive APs in the train is then,

$$R_{n+1} = 1 + (R_n - R_n u_n - 1)\exp\left(\frac{-\Delta t_n}{\tau_{rec}}\right), \tag{A2}$$

where Δt is the time interval between nth and $(n+1)$th AP and where,

$$u_{n+1} = U + u_n(1 - U)\exp\left(\frac{-\Delta t_n}{\tau_{facil}}\right). \tag{A3}$$

The synaptic response that is generated by any AP in a train is therefore given by,

$$EPSP_n = A \cdot R_n \cdot u_n. \tag{A4}$$

Synaptic connections displaying depression are characterized by negligible values of τ_{facil} and hence $u_n = U$.

Appendix G: Synaptic Classes

Current experiments indicate that two categories of synapse types exist (depressing and facilitating) and that at least four canonical classes of facilitating synapses, each with two subclasses, exist (Wang et al., 1999). Each class has three characteristic features:

Table A.1. Characteristic features of synaptic classes.

Features	1.	2.	3.
Description	**Low p (such that facilitation is observed for the second response in a train at any frequency).**	$\tau_{rec} \ll \tau_{facil}$	$\tau_{facil} > 1$ second
Class F1A	Yes	Yes	Yes
Class F1B	Yes but p is high (such that facilitation is not evident for most frequencies; present only at very low frequencies).	Yes	Yes
Class F2A	Yes	$\tau_{rec} \gg \tau_{facil}$	$\tau_{rec} > 0.5s.$
Class F2B	Yes but p is high.	$\tau_{rec} \gg \tau_{facil}$	$\tau_{rec} > 0.5s.$
Class F3A	Yes	$\tau_{rec} == \tau_{facil}$	τ_{rec} and $\tau_{facil} < 0.5s.$
Class F3B	Yes but p is high.	$\tau_{rec} == \tau_{facil}$	τ_{rec} and $\tau_{facil} < 0.5s.$
Class F4A	Yes	$\tau_{rec} == \tau_{facil}$	τ_{rec} and $\tau_{facil} > 0.5s.$
Class F4B	Yes but p is high.	$\tau_{rec} == \tau_{facil}$	τ_{rec} and $\tau_{facil} > 0.5s.$

Appendix H: Paired Pulses

This is another way of saying that paired pulse responses are not sufficient to characterize a synapse. Paired pulse protocols have been used extensively since the studies of Katz (del Castillo and Katz, 1954), even though this protocol leads to inconclusive and misleading results; paired synaptic responses are not sufficient to capture the interplay between release, depression and facilitation. Paired pulse ratios have been

used extensively to determine whether synaptic modifications are pre- or postsynapti-
cally expressed. However, since even facilitation has to be "allowed" for by the
kinetics of postsynaptic receptors one cannot make a conclusive statement about
changes in the ratio (Response1/Response2) (i.e. more receptors can change the ratio
if receptors where limiting), unless one can prove that receptors are not limiting.
Even if one can prove this, for many synaptic parameter configurations, changes in p,
τ_{rec} or τ_{facil} will appear uniform for the first few APs in a train with differential
changes only observed later (i.e. increased p can appear as an increase in receptors).
The paired pulse protocol is therefore the poorest possible measure of synaptic
responses and it should not be used at all.

Appendix I: Digitization of Synaptic Parameters

The precision of TI depends on the precision with which synapses can "record" the
timing of APs. This depends on whether the synaptic parameters are set in an ana-
logue or digital manner. In other words, if p has only two possible values and if synapses recover or facilitate in jumps then small differences in AP timing would be
digitized. The tolerance for sloppiness in timing would then depend on the extent of
digitization of synaptic parameters.

Appendix J: Steady State

At steady state beyond λ Hz, the synaptic response loses dependency on U (i.e. probability of release is no longer important in determining the average size of the synaptic response);

$$EPSP_{st} \approx \frac{A}{r \cdot \tau_{rec}}. \qquad (A5)$$

The limiting frequency is a characteristic frequency of synapses since it reflects a
specific relationship between neurotransmitter release and recovery from depression
and, if present, facilitation (Tsodyks and Markram, 1997). For depressing synapses;

$$\lambda \approx \frac{1}{U \cdot \tau_{rec}}. \qquad (A6)$$

Facilitating synapses also obey the $1/f$ rule, but at much higher frequencies. The
formulation to determine λ for facilitating synapses is more complex (see Tsodyks et
al., 1998). Typical values of λ are 5–30 Hz for depressing synapses and 70–130 Hz
for facilitating synapses.

The peak frequency (θ) is the frequency at which facilitating synapses produce the largest response. The relationship between θ and the model parameters can be approximated for any facilitating synapse:

$$\theta \approx \frac{1}{\sqrt{U \cdot \tau_{rec} \cdot \tau_{facil}}}. \tag{A7}$$

Typical values of θ range from 3 to 30 Hz for connections from neocortical synapses (Markram et al., 1998a). These characteristic parameters of frequency-dependent synapses can be used to quantify the impact of synaptic modifications in terms of changes in λ or θ. Changes in λ or θ in turn allows quantification of the impact of synaptic modifications on the information transmission capacity of synapses.

The steady state value of R (R_{st}) for a given frequency (r) of stimulation is given by:

$$R_{st}(r) = \frac{1 - \exp\left(\frac{-1}{r\tau_{rec}}\right)}{1 - (1 - u_{st}(r))\exp\left(\frac{-1}{r\tau_{rec}}\right)}, \tag{A8}$$

where

$$u_{st}(r) = \frac{U}{1 - (1 - U)\exp\left(\frac{-1}{r\tau_{facil}}\right)}. \tag{A9}$$

Appendix K: Inhibitory Synapses

In recent studies we found (Wang et al., 1999) that the synaptic properties of neocortical inhibitory synapses are highly heterogeneous and that they could be characterized according to six of the classes specified (Appendix 7: Synaptic Classes). This indicates that inhibition is not imposed according to some average level of activity in a "balancing act" of excitation and inhibition, but according to complex spatio-temporal patterns of activity; i.e. specific patterns are enhanced and specific patterns are inhibited. Furthermore, inhibitory interneurons are embedded within the recurrent microcircuitry where they may be reciprocally connected to inhibitory interneurons and excitatory neurons and the effect of raising or lowering activity levels in inhibitory interneurons is not simply predictable.

Appendix L: Lack of Boundaries

The divergence factor (D_f) is the fraction of neurons projected to which do not project back while the recurrence factor (R_f) is the inverse. If one assumes a R_f of 0.1 and that each neuron projects to and receives from 1000 neurons, then each neuron in the network could contact and receive indirect input from each neuron in a network of 10^{14} neurons within 6–7 synaptic junctions (average around 3–4!). If one considers that the average axonal, synaptic and dendritic delays as well as the time to peak of synaptic responses is in the 2–6 ms range (see Markram et al., 1997a; Markram, 1997), then each neuron could influence or respond to any neuron in the network within 40 ms. This time would only rise by about 10 ms for R_f as high as 0.5 and be around 80ms if the R_f were 0.9. This indicates that there are no strict anatomical boundaries in the nervous system leaving each neuron under the potential influence of every other neuron.

Appendix M: Speed of RI Accumulation

In simulations of recurrent neural networks with nonlinear synapses, we tested how quickly recurrent feedback can begin to exert an effect on the discharge of neurons initially driven by a stimulus. The activity in a set of neurons was measured after applying a stimulus to a group of neurons in a network that was either disconnected or connected. The discharge pattern began to deviate in the connected (recurrent) network from the activity in the disconnected network in less than 5 ms and after 20 ms the activity patterns were entirely different.

Appendix N: Network Efficiency

A concept of network efficiency has been developed earlier by Barlow in which it was important that objects be represented by as few discharging neurons as possible (see Martin, 1994 for a perspective). Another way to consider the efficiency of neural network activity is to consider directly what happens to the synaptic responses during a stimulus. When a stimulus is applied to a network of neurons it seems inevitable that some of the neurons will discharge at times when their synaptic responses would arrive at inappropriate times on the target neurons, for example during the refractory period or when the neuron is being inhibited; i.e. when the target neuron is unable to utilize the synaptic input maximally. At stimulus onset, therefore, it would appear that a large fraction of synaptic responses could be wasted because of mal-aligned neuronal activities. Due to recurrence, the per AP synaptic impact could rise progressively until a minimal number of AP discharges would produce a maximal response.

Appendix O: The Binding Problem of the Binding Problem

The binding problem relates to how features of objects are bound to provide a single percept of an object (see von der Malsburg, 1986; Singer, 1993; Singer and Gray, 1995). The binding problem as stated implicitly assumes that objects are isolated from other objects. The problem with the binding problem as stated, therefore, is that it does not address what binds different objects, different images, different percepts or entire chunks of different events and does not address what represents "empty space".

References

Abbott, L.F., Varela, J.A., Sen, K., Nelson, S.B. (1997) Synaptic depression and cortical gain control. Science 275: 220–224.

Ales, E., Poyato, J.M., Valero, V., Alvarez de-Toledo, G. (1998) Neurotransmitter release: A process of membrane fusion occurring in fractions of milliseconds. Rev Neurol. 27: 111–117.

Atwood, H.L., Wojtowicz, J.M. (1986) Short-term and long-term plasticity and physiological differentiation of crustacean motor synapses. Int Rev Neurobiol 28: 275–362.

Barlow, H.B. (1961) Possible principles underlying the transformation of sensory messages. In: W. Rosenblith (Ed.) Sensory Communication. Cambridge, MA: MIT Press, pp. 217–234.

Betz, W.J. (1970) Depression of transmitter release at the neuromuscular junction. J Physiol (Lond) 206: 626–644.

Brodin, L., Low, P., Gad, H., Gustafsson, J., Pieribone, V.A., Shupliakov, O. (1997) Sustained neurotransmitter release: New molecular clues. Eur J Neurosci 9: 2503–2511.

Brodin, L., Shupliakov, O. (1994) Functional diversity of central glutamate synapses: pre- and post-synaptic mechanisms. Acta Physiol Scand 150: 1–10.

Byrne, J.H. (1982) Analysis of synaptic depression contributing to habituation of gill-withdrawal reflex in Aplysia californica. J Neurophysiol 48: 432–438.

Carpenter, G.A., Grossberg, S. (1981) Adaptation and transmitter gating in vertebrate photoreceptors. Journal of Theoretical Neurobiology 1: 1–42.

Carpenter, G.A., Grossberg, S. (1983) A neural theory of circadian rhythms: the gated pacemaker. Biological Cybernetics 48: 35–59.

Curtis, D.R., Eccles, J.C. (1960) Synaptic action during and after repetitive stimulation. J Physiol (Lond) 150: 374–398.

Cuvier, G. (1817) Le Regne Animal Distribue d'apres son Organisation. Paris: Deterville.

del Castillo, J., Katz, B. (1954) Statistical factors involved in the neuromuscular facilitation and depression. J Physiol (Lond) 124: 574–585.

Dityatev, A.E., Clamann, H.P. (1998) Synaptic differentiation of single descending fibers studied by triple intracellular recording in the frog spinal cord. J Neurophysiol 79: 763–768.

Dudel, J. (1965) Potential changes in the crayfish motornerve terminal during repetitive stimulation. Pflügers Arch gen Physiol 282: 323–337.

Eccles, J.C. (1946) Synaptic potentials of motoneurones. J Neurophysiol. 9: 87–120.

Eccles, J.C. (1964) The Physiology of Synapses. Berlin: Springer-Verlag OHG.

Eccles, J.C., Katz, B., Kuffler, S.W. (1941) Nature of the "end-plate" potential in curarized muscle. J Neurophysiol 4: 362–387.

Feng, T.P. (1940) Studies on the neuromuscular junction. XXVIII. The local potentials around N-M junctions induced by single and multiple volleyes. Chin J Physiol 16: 341–372.

Francis, G., Grossberg, S. (1996) Cortical dynamics of form and motion integration: Persistence, apparent motion, and illusory contours. Vision Research 36: 149–173.

Francis, G., Grossberg, S., Mingolla, E. (1994) Cortical dynamics of feature binding and reset: Control of visual persistence. Vision Research 34: 1089–1104.

Frost, W.N., Katz, P.S. (1996) Single neuron control over a complex motor program. Proc Natl Acad Sci USA 93: 422–426.

Gerrad, R. (1949) Physiology and psychiatry 106: 161–173.

Grossberg, S. (1968) Some physiological and biochemical consequences of psychological postulates. Proceedings of the National Academy of Sciences 60: 758–765.

Grossberg, S. (1969a) On the production and release of chemical transmitters and related topics in cellular control. J Theor Biol 22: 325–364.

Grossberg, S. (1969b) On learning and energy-entropy dependence in recurrent and nonrecurrent signed networks. Journal of Statistical Physics 1: 319–350.

Grossberg, S. (1972) A neural theory of punishment and avoidance, II: Quantitative theory. Mathematical Biosciences 15: 253–285.

Grossberg, S. (1976) Adaptive pattern classification and universal recoding, II: Feedback, expectation, olfaction, illusions. Biological Cybernetics 23: 187–202.

Grossberg, S. (1980) How does a brain build a cognitive code? Psychological Review 87: 1–51.

Grossberg, S. (1984a) Some normal and abnormal behavioral syndromes due to transmitter gating of opponent processes. Biological Psychiatry 19: 1075–1118.

Grossberg, S. (1984b) Some psychophysiological and pharmacological correlates of a developmental, cognitive, and motivational theory. In: R. Karrer, J. Cohen, P. Tueting (Eds.) Brain and Information: Event Related Potentials. New York: New York Academy of Sciences, pp. 58–142.

Grossberg, S. (1992) Cortical dynamics of visual motion perception: Short-range and long-range apparent motion. Psychological Review 99(1): 78–121.

Grossberg, S., Merrill, J.W.L. (1992) A neural network model of adaptively timed reinforcement learning and hippocampal dynamics. Cognitive Brain Research 1: 3–38.

Grossman, Y., Spira, M.E., Parnas, I. (1973) Differential flow of information into branches of a single axon. Brain Res 64: 379–386.

Henkel, A.W., Almers, W. (1996) Fast steps in exocytosis and endocytosis studied by capacitance measurements in endocrine cells. Curr Opin Neurobiol 6: 350–357.

Kamiya, H., Zucker, R.S. (1994) Residual Ca^{2+} and short-term synaptic plasticity. Nature 371: 603–606.

Korn, H., Faber, D.S., Triller, A. (1986) Probabilistic determination of synaptic strength. J Neurophysiol 55: 402–421.

Liaw, J.S., Berger, T.W. (1996) Dynamic synapse: A new concept of neural representation and computation. Hippocampus 6: 591–600.

Liley, A.W., North, K.A.K. (1953) An electrical investigation of the effects of repetitive stimulation on mammalian neuromuscular junction. J Neurophysiol 16: 509–527.

Lloyd, D.P.C. (1949) Post-tetanic potentiation of responses of monosynaptic reflex pathways of the spinal chord. J Gen Physiol 33: 147–170.

Magleby, K.L. (1987) Short-term changes in synaptic efficacy. In: G.M. Edelman, W.E Gall, W.M. Cowman (Eds.) Synaptic Function. New York: Wiley, pp. 21–56.

Magleby, K.L., Zengel, J.E. (1982) A quantitative description of stimulation-induced changes in transmitter release at the frog neuromuscular junction. J Gen Physiol 80: 613–638.

Mallart, A., Martin, A.R. (1967) An analysis of facilitation of transmitter release at the neuromuscular junction of the frog. J Physiol (Lond) 193: 679–694.

Markram, H. (1997) A network of tufted layer 5 pyramidal neurons. Cerebral Cortex 7: 523–533.

Markram, H., Lübke, J., Frotscher, M., Roth, A., Sakmann, B. (1997a) Physiology and anatomy of synaptic connections between thick tufted pyramidal neurones in the developing rat neocortex. J Physiol (Lond) 500: 409–440.

Markram, H., Lübke, J., Frotscher, M., Sakmann, B (1997b) Regulation of synaptic efficacy by coincidence of postsynaptic APs and EPSPs. Science 275: 213–215.

Markram, H., Wang, Y., Tsodyks, M. (1998a) Differential signaling via the same axon of neocortical pyramidal neurons. Proc Natl Acad Sci USA 95: 5323–5328.

Markram, H., Pikus, D., Gupta, A., Tsodyks, M. (1998b) Potential for multiple mechanisms, phenomena and algorithms for synaptic plasticity at single synapses. Neuro Pharmacol 37: 489–500.

Markram, H., Gupta, A., Uziel, A., Yung, W., Tsodyks, M. (1998c) Information processing with frequency dependent synapses. Neurobiol Learning and Memory 70: 101–112.

Markram, H., Tsodyks, M. (1996). Redistribution of synaptic efficacy between neocortical pyramidal neurons. Nature 382: 807–810.

Martin, K.A.C (1994) A brief history of the "Feature Detector". Cerebral Cortex 4: 1–7.

Murthy, V.N., Stevens, C.F. (1998) Synaptic vesicles retain their identity through the endocytic cycle. Nature 392: 497–501.

Parnas, H., Parnas, I. (1994) Neurotransmitter release at fast synapses. J Membr Biol 142: 267–279.

Parnas, I. (1972) Differential block at high frequency of branches of a single axon innervating two muscles. J Neurophysiol 35: 903–914.

Pevsner, J., Scheller, R.H. (1994) Mechanisms of vesicle docking and fusion: Insights from the nervous system. Curr Opin Cell Biol 6: 555–560.

Pongs, O. et al., (1993) Frequenin: a novel calcium-binding protein that modulates synaptic efficacy in the Drosophila nervous system. Neuron 11: 15–28.

Senn, W., Tsodyks, M., Markram, H. (1997) An algorithm for synaptic modification based on precise timing of pre- and postsynaptic APs. Lecture Notes in Computational Science 97: 121–126.

Sherrington, C.S. (1897) The central nervous system, Vol. III. In: M. Foster (Ed.) A Textbook of Physiology, 7th Edition. London: Macmillan.

Shupliakov, O., Low, P., Grabs, D., Gad, H., Chen, H., David, C., Takei, K., DeCamilli, P., Brodin, L. (1997) Synaptic vesicle endocytosis impaired by disruption of dynamin-SH3 domain interactions. Science 276: 259–263.

Singer, W. (1993) Synchronization of cortical activity and its putative role in information processing and learning. Annu Rev Physiol 55: 349–374.

Singer, W., Gray, C.M. (1995) Visual feature integration and the temporal correlation hypothesis. Annu Rev Neurosci 18: 555–586.

Softky, W.R., Koch, C. (1993) The highly irregular firing of cortical cells is inconsistent with temporal integration of random EPSPs. J Neurosci 13: 334–350.

Somogyi, P., Tamas, G., Lujan, R., Buhl, E.H. (1998) Salient features of synaptic organization in the cerebral cortex. Brain Research Rev 26: 113–135.

Sperry, R. (1966) Brain bisection and mechanisms of consciousness. In: J.C. Eccles (Ed.) Brain and Conscious Experience. New York: Springer-Verlag.

Thomson A.M. (2000) Molecular frequency filters at central synapses. Prog Neurobiol 2000 62(2): 159–196.

Thomson, A.M., Deuchars, J. (1994) Temporal and spatial properties of local circuits in neocortex. TINS 17: 119–126.

Thomson, A.M., Deuchars, J. (1997) Synaptic interactions in neocortical local circuits: Dual intracellular recordings in vitro. Cereb Cortex 7: 510–522.

Tsodyks, M., Markram, H. (1996) Plasticity of neocortical synapses enables transitions between rate and temporal coding. Lecture Notes in Computational Science 96: 445–450.

Tsodyks, M., Markram, H. (1997) The neural code between neocortical pyramidal neurons depends on neurotransmitter release probability. Proc Natl Acad Sci USA 94: 719–723.

Tsodyks, M, Pawelzik, K., Markram, H. (1998) Neural networks with dynamic synapses. Neural Computation 10: 821–835.

von der Malsburg, C. (1986) Am I thinking assemblies. In: G. Palm, A. Aertsen (Eds.) Brain Theory. Berlin: Springer-Verlag.

von der Malsburg, C., Bienenstock, E. (1986) Statistical coding and short-term synaptic plasticity: A scheme for knowledge representation in the brain. In: E. Bienenstock, F. Fogelman-Soulie, G. Weisbuch (Eds.) Disordered Systems and Biological Organization. Berlin: Springer-Verlag.

Wiersma, C.A.G., Ikeda, K. (1964) Interneurons commanding swimmeret movements in the crayfish, Procambarus clarki (Girard). Comp Biochem Physiol 12: 509–525.

Wang, Y., Gupta, A., Markram, H. (1999) Anatomical and functional differentiation of glutamatergic synaptic innervation in the neocortex. Journal of Physiology (Paris) 93(4): 305–317.

Zucker, R. S. (1989) Short-term synaptic plasticity. Ann Rev Neurosci 12: 13–31.

Chapter 6

The Development of Cortical Models to Enable Neural-based Cognitive Architectures

Thomas McKenna

6.1 Introduction

The development of models of the cerebral cortex parallels the growth in sophistication of neural models in general, proceeding from heuristic models, to functional "black box" systems approaches, to large-scale neural circuit models. Currently, computational neuroscience is producing increasingly detailed neurobiologically based models of cortex and related structures. Many of these are intended as simulations of the biology, and are subject only to the constraint of predicting experimental observations in the neurobiological domain. A few of these models have some computational capability in the general domains of pattern recognition or control (Ambros-Ingerson et al., 1990; Carpenter and Grossberg, 1991; McKenna, 1994; Zornetzer et al., 1995).

On the other hand, cognitive scientists have developed increasingly more elaborate cognitive architectures with computational capabilities for a wide range of tasks that, in humans, require cognitive skills. These cognitive architectures, however, have not exploited our current understanding of cortical neurobiology. My goal is to characterize this gap between the biologically realistic cortical models, with very limited executive capability, and cognitive architectures, based on Artificial Intelligence (AI) software formalisms, which have a range of "cognitive" capabilities, but are not autonomous, and have yet to exploit emerging neurobiological principles. I will point to several promising new directions in computational and cognitive neuroscience which might help bridge this enormous chasm. It is hoped that computational neuroscientists will take seriously the need to redefine the goals of developing cortical models in terms of the broader aim of accounting for cognitive skills rather than continuing to limit their scope to the simulation of strictly biological processes or mechanisms of sensory phenomena.

6.1.1   Computational Neuroscience Paradigms and Predictions

Neuroscience developed over the second half of this century as a fusion of traditional approaches (disciplines) including anatomy, physiology, biochemistry, pharmacology, and physiological psychology. As neuroanatomical and neurophysiological techniques improved, a microarchitecture of cerebral cortex began to emerge that was very mechanistic, consisting of a synaptology of cell types. The successes of Eccles and his associates in characterizing spinal networks of excitatory and inhibitory neurons permitted Lorente de No and Szentagothai to build elaborate circuit diagrams of the cerebral cortex. These diagrams, however, were static representations and could not account for any cortical functions, as defined by the long tradition of neurology, neuroanatomy and neuropsychology of cortex. In fact these diagrams were only marginally more informative than those of Ramon y Cajal which were built entirely from anatomical data (DeFelipe and Jones, 1988). Such diagrams, however, were sufficient to stimulate many early cortical modeling efforts, including the seminal work of McCulloch and Pitts on general models of neural computation.

The effective use of computer simulations to implement realistic simulations of cortex begins with the work of Rall, Perkel and Shepard in which neuronal geometry and biophysics could be represented as computationally tractable single neuron models that could be connected into large networks (Segev et al., 1995). This process has continued with development of computational neuroscience, a fusion of neuroscience and computer science. A number of software tools, and simulation environments have been produced that permit a high degree of biological realism in neural models (e.g. Bowers and Beeman, 1998). Large-scale models of neural systems, such as cortical simulations, have been implemented which can generate dynamic patterns of activity, but brain imaging technology is just beginning to achieve the temporal and spatial resolution for validation or interpretation of the spatio-temporal patterns emerging in the simulations.

There is, however, sufficient data on the discharge patterns, receptive fields, and task relevant discharge of single cortical neurons to enable the use of single unit properties as a constraint and prediction of the models. Hence, many of the current large-scale models of cortex are based on sensory cortex, in particular visual cortex, and seek to predict either sensory receptive field properties or emergent perceptual behavior, such as visual illusions (Grossberg, 1997). Given the (reductionist) neuroscience view that neural models should be based on knowledge of the underlying cortical functional organization, the relative paucity of such data in associative cortex compared with primary visual cortex has resulted in fewer attempts at modeling cortical areas most directly implicated in cognitive skills, such as the frontal cortex, except for a few heuristic models or connectionist models (several promising efforts are discussed below).

As human brain imaging data emerges that links patterns of brain activation to human performance, or in the best cases, critically test theories of cognitive architecture, it will provide important new challenges and opportunities to those modeling cerebral cortex. One such trend is described in Chapter 8 by Taylor in this volume. This is the attempt to produce structural models that account for the sequential and co-activation patterns of different cortical regions during tasks that involve cognitive skills. This is the level at which connectionist models may suffice to account for specific observations in the brain imaging data, without having general cognitive executive capability.

However, as the spatial and temporal resolution of brain imaging improves, more details of the dynamics of the activation patterns are revealed, at what I will refer to as the cortical mesoscale. This is a level of functional organization below a cortical region (e.g. Brodmann's areas) but beyond the cortical cell column or the region sampled by multiunit recording from a single micro-electrode. Data at the mesoscale will provide opportunities and challenges to cortical models. Indeed, successful cortical modeling of the dynamics at this level may have to await a theory of cortical dynamics grounded in microcircuitry. One significant early attempt to account for quantitative features of higher-resolution brain imaging during cognitive tasks, using models with executive capability is provided by the work of Carpenter and Just (discussed below).

There currently are no computational neuroscience based models of cortex that address higher cognitive skills. In fact quite a range of cognitive phenomena are simply not being addressed by computational neuroscience. However, there are very promising efforts underway that are developing some key elements that will enable neural-based cognitive architectures. First, biologically realistic cortical models are now being developed which have the scale, learning algorithms, and architectures to tackle challenging, high-level pattern recognition. Second, connectionist architectures motivated by the modeling of cognitive skills are incorporating more biological principles, resulting in more autonomous capability. I think it is important that these approaches be fused in the next generation of cortical models.

6.2   The Challenge of Cognitive Architectures

6.2.1   General Cognitive Skills

Before reviewing cognitive architectures, one needs to describe what is encompassed by cognitive skills, since these are what cognitive architectures are intended to explain or mimic. A list of general cognitive skills might include:

1.   learning and knowledge acquisition;

2.   use of a working memory to hold multiple outcomes for assessment;

3.   situation assessment;

4.   creating and maintaining goal hierarchies;

5.   generating multiple plans;

6.   predicting consequences of actions (e.g., judgement);

7.   self management and control;

8.   reading and writing;

9.   language understanding;

10. math/logic;

11. problem solving (including the special case of scientific problem solving);

12. communication;

13. teaching (modified after Chipman, 1992).

Similar lists appear in the works of Newell (1992) and Anderson (1993). The first three skills, focused on learning and memory capabilities, have been addressed by a series of modeling efforts, beginning with mathematical psychology models that served as the forerunner of associative memory models and PDP models, through the connectionist networks, and more recently by computational cognitive neuroscience models focused on learning mechanisms in hippocampal and paleocortical systems. The fourth topic, situation assessment, has been addressed to some extent by artificial neural networks and hybrid neural networks as extensions of pattern recognition technology, but a true context-sensitive situation assessment still requires a significant AI component. Even these hybrid systems are restricted to well-characterized task domains. Some language skills have been addressed by semantic network models and the results critically tested with human brain imaging data, but the goal of simulating language processing is still elusive. Moreover, the semantic networks rely on hand-crafted linguistic features, and are not extensible to natural language. All of the listed skills have been addressed to some extent by production-based cognitive architectures (see below). The challenge and opportunity for computational cognitive neuroscience at this stage is to develop models capable of the skills of planning, goal maintenance and cognitive control (skills 5–8 in preceding list) yet retaining links to functional sensori-motor systems and the systems supporting learning and memory that would enable an autonomous cognitive architecture that could operate directly on the world.

6.2.2   A Survey of Current Cognitive Architectures

I think it is important for those developing computational models of cortex to be aware of the cognitive architectures that have emerged from cognitive science and AI to model cognitive skills. I will here summarize the major cognitive architectures for the benefit of neuroscientists, with the caveat that I view them from the perspective of computational neuroscience rather than cognitive science.

6.2.2.1   Soar

Soar is a software system originally intended to be an artificially intelligent software architecture with psychological inspiration (Newell, 1992; Rosenbloom et al., 1993). It is actually a project that has generated successive iterations of a software architecture. It has a dual purpose, the first as an architecture for integrated intelligent systems, and the second being the development of a cognitive architecture that models human cognition. After its initial development, Soar was recast by Allen Newell as a unified theory of cognition. Soar is implemented as an architecture, where architecture refers to a software architecture having a fixed set of mechanisms that enable acquisition and use of content in a memory to guide behavior in the pursuit of goals. Soar is basically a symbol system, but relies on a tightly coupled hierarchy of layers in its knowledge-level structure. These levels are memories, tasks and decisions set by the "environment", and the goals of the system. The levels form a hierarchy of nested loops. Soar is a computational problem-solving system organized around defined problem spaces and production systems. These production systems have a long-term memory (LTM) for programs and data consisting of parallel condition–action rules. A more mechanistic account is provided by consideration of the three major components of Soar:

1.   a long-term recognition memory storing multiple productions which passes output to

2.   a working memory for current problem spaces containing a goal stack with output to

3.   a chunking module (Figure 6.1).

Within the goal stack a problem proceeds until it meets an impasse at which point a sub-goal is generated, etc. The chunking module constructs new productions (chunks) to capture the knowledge developed in working memory, resulting in recognition memory. Hence the system converts problem solving into recognition of known problems. This results in a "recognize–act" cycle. Soar does not have a sub-symbolic level. Newell has presented some ad hoc arguments against human neural sub-symbolic operations. He argues that in humans about 100 msec long subsystem operations (retrieval cycle including recognition memory and decision) underlie cognition and that thereby simple neural circuits (which take 10 msec to operate) could not possibly be the substrate of these operations, since they would have too few cycles to operate. This of course assumes a serial operation of neural circuit transactions. Recent neuroscience and connectionist results on synchronous neural computation, spike-based computation, and nonlinear neural systems dynamics provide alternatives to this serial view (which ought to be exploited in a neural-based cognitive architecture, see below). As a cognitive architecture modeling human cognition, Soar has been criticized on several grounds. First, Soar has a working memory with around 3,000 items, clearly well beyond human capacities. Second, the chunking operation in Soar combines large numbers of elements in complicated operations, whereas in humans about three items are typically chunked. Thirdly, Soar productions are much more fine-grained than most psychological descriptions of cognitive tasks.

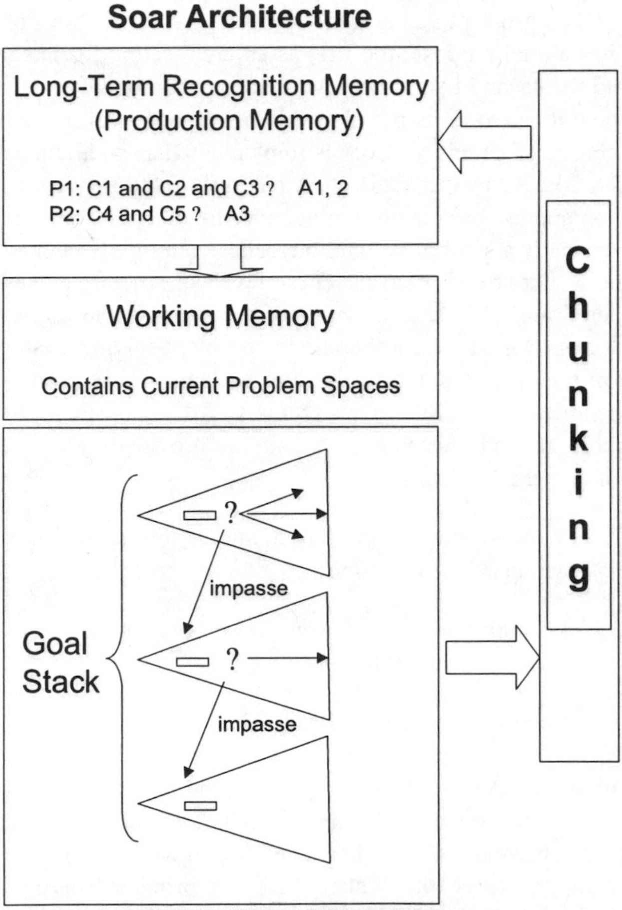

Figure 6.1. The Soar Cognitive Architecture (P: Productions, C: Conditions, A: Actions). Adapted from Newell (1992: 25–79, Figure 2) with kind permission of Kluwer Academic Publishers.

In order to handle high level perceptual and motor skills, Soar has been combined with the general human-computer interaction formalism GOMS (Goals, Operators,

Methods, and Selection rules). The emphasis has been on high level tasks in this domain, since the sensory and motor systems that have been integrated with Soar to date are somewhat primitive and unrealistic. On the other hand, Soar has made predictions in concert with human cognition including:

1. the reconstructive nature of declarative memory;

2. the fast read/slow write feature of human memory;

3. the fact that short-term and long-term goal structures are independent;

4. the observation that learning rate is tied to the impasse rate. Human cognitive skills which have been modeled with Soar include calculator use, automobile driving, computer game playing, puzzle solving, and aspects of language learning.

Further critical discussions of Soar can be found in Feldman (1991), Vincente and Kirklik (1992), Chandrasekaran (1994), and Cooper and Shallice (1995).

6.2.2.2   ACT-R

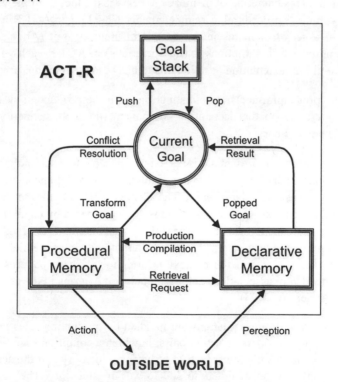

Figure 6.2. The ACT-R Cognitive Architecture. Adapted with permission from Anderson (1993).

The ACT-R (Adaptive Control of Thought-Rational) theory and model is predicated on the view that cognitive skills are realized by production rules and that these rules are modular pieces of knowledge underlying steps in cognition (Anderson, 1993). ACT-R is then a theory of how these production rules are executed. In this theory, complex cognitive processes are achieved by stringing together a sequence of production rules by appropriate setting of goals and writing-to and reading-from working memory. The production rules are abstract in that they are intended to apply to many situations. The production system operations include a pattern matching between the conditions of the production and contents of working memory, conflict resolution between alternative production rules, and execution or firing of the production rule performed (Figure 6.2). ACT-R has a neural-like implementation, incorporating a number of mechanisms from artificial neural nets and parallel models of associative memory. ACT-R is "activation based", e.g. declarative memory structures vary in the level of activation, and the rate of production-rule matching is set by the activation level of declarative structures. Learning involves increasing the strength of the declarative structures and production rules. While the implementation is neural-net, the system is symbolic at the algorithm level. In ACT-R, chunking of working-memory elements (WMEs) takes place in a hierarchical organization, but the basic chunking operation involves a small number of elements. The WMEs are considered a cognitive unit. This chunking of memory elements is distinct from the chunking in Soar which combines procedural learning elements. Chunking in ACT-R is the means of organizing a set of elements into a long-term memory unit (Anderson, 1993). There are a number of assumptions (basic operations) in ACT-R each with a corresponding neural implementation or rationalization. These assumptions are:

1. A procedural-declarative distinction, with the production system operating on the declarative component (no such distinction is made in Soar).

2. Declarative representations can be composed into a set of chunks, where each chunk consists of a limited number of elements (about 3) in specific relationships. The elements of a chunk are themselves chunks. Complex structures are represented by chunk hierarchies. In terms of its neural implementation, a chunk is represented as a pattern of activation that associates the elements of the chunk and the context information. "These patterns are encoded in associations among modules of neurons in the cortex" (Anderson, 1993).

3. Procedural representations are handled by production rules organized into condition-action pairs, in which a condition specifies a set of things that must be true of working memory and the action specifies a set of things to add to working memory. The neural implementation of the productions are modules for holding variable bindings that define the production instantiation. Control ele-

ments sequence the matching of chunks and broadcast to declarative memory the chunks created in the action.

4. Goals are ordered in LIFO goal stacks, with the top goal being the source of activation. Goals can have specific utilities or values associated with them, and every production that fires must respond to some goal on some stack. In the neural implementation goal stacks can be encoded by pairwise associations of goal chunks in declarative memory. "Hierarchical goal processing is implemented by the prefrontal cortex" (Anderson, 1993).

5. The top elements on goal stacks and elements focused on data coming from the environment (including internal) are sources of activation. When a goal is popped, it is removed as a source of activation. In the neural implementation, elements being "perceived" are sources of activation. Patterns representing these elements are superimposed in the context mode within declarative memory.

6. Associated with chunks are activation levels (within neural modules) that reflect the log odds that they will match a particular production instantiation. The base-level activation of a chunk is combined with activation from associated sources, using weighted sums in its neural implementation.

7. The chunks in a production condition are matched sequentially, and the total time to calculate the instantiation to production is summed over all chunks matched in the condition, but decays exponentially with a rate dependent on both the activation and production strength.

8. The utility of productions are evaluated at match based on P (probability of reaching a goal), G (the value of the goal) and C (the expected cost of the goal), calculated as PG–C implemented as positive and inhibitory transactions.

9. Chunks have base levels of activation reflecting their log prior odds of matching a chosen instantiation. Associations among chunks have strengths that reflect the log likelihood ratios of one chunk being a source for the chunk that is going to match the instantiation. "There appears to be a significant role of the hippocampus in building up strength in declarative memory. Long-term potentiation appears to follow the predictions of (the) equation" used to calculate "the base level of activation of a chunk" (Anderson, 1993).

10. Production strength, a measure of the log odds that an instantiation of the production will fire. Productions acquire strength according to the same equation used in (9) where a production is defined as a firing of a production instantiation. In the neural implementation, production strength helps resolve contention for access to declarative memory.

The ACT-R system continues to evolve with the help of a vigorous user community. ACT-R has been applied to modeling a number of cognitive skills including: visual search (Anderson et al., 1997); task switching (Sohn et al., 2000); driving behavior (Salvucci et al., 2002); cognitive arithmetic (Labiere, 1998); learning by exploration and demonstration (Cox and Young, 2000); problem solving (Gluck, 2000; Gunzelmann and Anderson, 2001); spatial reasoning (Schunn and Harrison, 2001); and insight and scientific discovery (Schunn and Anderson, 1998).

6.2.2.3   EPIC

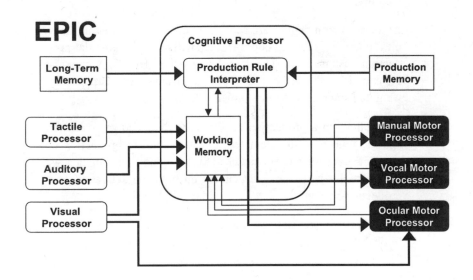

Figure 6.3. The EPIC Cognitive Architecture. From Meyer and Kieras (1997). © 1997 American Psychological Association. Adapted with permission.

EPIC (Explicit representation of Executive cognitive Processes that Interact with other processes to Control performance) is a computational framework and an architecture for constructing models of human performance which subsumes many results and theories from the human performance literature. It includes both cognitive and perceptual/motor processes, and is realized as a software system for simulation modeling (Meyer and Kieras, 1997). The basic assumptions are the use of a production rule for the cognitive processor and parallel perceptual and motor processors (Figure

6.3). The components, pathways and most time parameters are fixed properties. However, the cognitive processor production rules, the perceptual recoding and the response requirements are task-dependent properties.

The perceptual processors (auditory, visual, and tactile) operate on symbolically coded packages of features rather than raw sensor data, and the outputs are items placed in modality-specific partitions of working memory. For example, the auditory module detects onsets and offsets, encodes tones and sounds, and outputs speech input as a temporally chained representation that decays with time. The visual module contains an "eye model" that represents "visual properties" depending on retinal zone, encodes perceptual properties of within a visual working memory, maintains an internal representation of visual objects and reports changes to an "ocular motor processor". The tactile module processes kinesthetic feedback from motor effectors that identify "movement states".

The motor processors receive symbolic instructions from the cognitive processor providing symbolic movement specifications and times. In the motor processing module, movement instructions are expanded into motor features including style, effector, direction and extent. The cognitive processor is informed of the current state. An ocular motor processor generates "eye movements" from commands or visual events. Both feature preparation time and execution time are included, and it is claimed that many eye movement phenomena can be accounted for. A manual motor processor controls both hands by use of a single processor, and a variety of hand movement styles can be produced including pointing, button pushing, typing and controlling.

The cognitive processor is programmed with production rules, and these rules represent all the procedural knowledge required to perform the task. The production system formalism employed is called Parsimonious Production System (PPS) and it has a working memory, production rules expressed as condition-action statements and a rule interpreter. PPS is parallel in the sense that multiple production rules are tested at the same time. EPIC has three functionally distinct memory stores:

1.   declarative long-term memory contains knowledge expressed as propositions;

2.   procedural memory containing PPS production rules that instantiate procedural knowledge for performing tasks;

3.   working memory contains symbolic control information needed for testing and applying the production rules stored in procedural memory. The working memory also stores symbolic representations of stimulus inputs and response outputs are also stored for use by the system's production rules. Although EPIC has memory it has no learning algorithms. Also, EPIC does not have an attentional spotlight, or internal filter. Internal cognitive structures are added to EPIC only as required by the task being modeled.

EPIC has been applied to the modeling of a number of human tasks, with a particular emphasis on predicting the timing of actions, revealing its roots in psychological theories of reaction time. Since EPIC employs stochastic elements (in distinction to ACT-R) it can model distributions of reaction times rather than mean reaction times. It has been applied to verbal working memory tasks, tracking/choice dual tasks, human-computer interaction, and telephone operator tasks.

A hybrid of ACT-R and EPIC has been developed called ACT-R/PM which incorporates a parallel set of perceptual-motor modules which enable it to model high performance tasks and make specific timing predictions.

6.2.2.4   CAPS

Just and Carpenter developed a computational modeling system called CAPS (Concurrent Activation-based Production System) initially to model reading tasks (Thibadeau et al., 1982). CAPS is a symbolic system with some connectionist properties: continuous activation levels and reiterative propagation of activation. CAPS was subsequently applied to modeling the processes in mental rotation (Just and Carpenter, 1985) and working memory processes engaged in performing the Raven Progressive Matrices Test (Just et al., 1990, see discussion in later section). Later the model was modified to 3CAPS (Capacity Constrained CAPS) (Just and Carpenter, 1992). In 3CAPS the total amount of activation in the system is contrained, and varied among "individuals", permitting the simulation of a wide range of individual differences in language comprehension. The most recent version of this model is 4CAPS (Cortical Capacity-Constrained CAPS) (Just et al., 1999). It was designed to be consistent with key properties of higher cortical function and they have explored the ability of this model to account for both the functional decomposition of the (human) cognitive system and levels of brain activation observed in fMRI imaging during sentence comprehension tasks (Just et al., 1999). As in the earlier versions, this architecture is essentially a production-system with connectionist properties. One of the significant innovations of 4CAPS is the use of a number of collaborating component computational systems, each of which is intended to correspond to the functions of a cortical area (e.g. Broca's area, Wernicke's area and dorsolateral prefrontal cortex). A significant design feature is that the graded activation within each component is a limited resource drawn on by computational activities, and the resource or capacity utilization rate in each component is intended to be comparable to the amount of brain activation observed in human brain imaging. Each component system is itself a production system, with procedural knowledge contained in an unordered set of if–then production rules, but with productions that fire reiteratively, over cycles, propagating activation from source elements to target elements, until they activate their action elements to threshold or something stops their firing. As processing proceeds, productions act like dynamic links associating conditions with actions. Multiple productions may collaborate, because on each cycle, productions whose enabling conditions are satisfied all fire, which may change the state for the next cycle. Just et al. (1999) have developed a corresponding theory of "collaboration with specialization" among cortical regions during cognitive task performance that enhances the value of their research for the development of a computational cognitive neuroscience.

6.2.3   Assumptions and Limitations of Current Cognitive Architectures

A common picture seen in current cognitive architectures is the progression of information from sensory system to working memory to motor system. This segmentation was largely abandoned by neuroscientists based on experimental results obtained over the past three decades. The picture of the CNS that has emerged has the following properties:

1. Neural plasticity is a broadly distributed property of the CNS, and is exhibited even within sensory systems (see e.g. Weinberger, 1995; Nicolelis et al., 1998).

2. The types of synaptic and neuronal plasticity and connectivity patterns vary across brain structures so that specific areas have special functions to contribute to learning and memory.

3. Sensory and motor representations are distributed across neural ensembles, and include spatio-temporally extended representations reflecting the actions of recurrent interactions that are both bottom-up and top-down (Nicolelis et al., 1995).

4. Many neuronal systems originally assigned attributes of either sensory or motor, are in fact sensori-motor, and participate in both functions. There are many neural structures (e.g. in the oculomotor system) where individual neurons exhibit both sensory receptive fields and participation in the coding of motor commands. Hence, for many sensorimotor control problems, intermediate processing by a separate, transforming neural system is not necessary, although other systems can modulate the transforms that take place within sensorimotor neurons and structures.

Aside from these basic architectural considerations, another important difference between the cognitive architectures and computational neuroscience models is that the former nearly always operate on refined data extracted as features or lexical items, whereas many neuroscience models explicitly operate on raw data from the world, e.g. complex but informative physical optical, acoustic, tactile and motion inputs. While it can be argued that this is because the cognitive architectures are addressing higher level skills, it also opens them to the critique that the models can only operate in a meta-world of higher sensory representations and will necessarily fail in autonomous cognition. In fact, the most serious critique of current cognitive architectures is the lack of a true active sensing in a complex 3D world, with all that that implies for exploring, adaptability, active representations, action planning, and all the other cognitive skills that animals and people exhibit when they are not sitting in front of a computer screen. Indeed, a factor that seriously limits the use of cognitive architectures in human-system design is the tremendous effort required to perform a detailed task analysis of such problems and to decompose them into elements upon which the cognitive architecture can operate. Hence the need for greater autonomy in cognitive architectures, which operationally implies both the ability to (learn how to) analyze the task and the ability to reconfigure its computational resources in a task-dependent manner. Reconfiguration of neural computational resources is a central theme in current computational neuroscience.

6.3   The Prospects for a Neural-based Cognitive Architecture

6.3.1   Limitations of Artificial Neural Networks

Artificial neural networks (ANNs) provide an alternative means of implementing the production-system cognitive architectures and are also the building block of connectionist architectures. Yet, current artificial neural nets have limited capabilities beyond that of pattern recognition and control applications. Hence it is not surprising that early attempts to mimic tasks like language processing or approximate reasoning have relied on hybrid architectures that combine neural networks with AI components. Many of the limitations of current ANNs derive from their being too simple and too homogeneous. One approach to overcoming this limitation is the use of modular heterogeneous architectures, that permit some degree of problem decomposition, while avoiding problems of interference from sequentially learned representations. However, deeper changes in neural network processing elements and architecture are needed to seriously raise the computational power of these networks. The stepwise introduction of biological neural principles (O'Reilly, 1998), and the exploration of recurrent, dynamic networks (Angeline et al., 1994) are promising developments to achieving greater computational power of neural networks. However, the real challenge and opportunity in neural systems is the identification and exploitation of the full range of neurobiological principles from neuron and synaptic dynamics, to micro-circuitry, ensemble encoding and dynamics, and global system dynamics.

6.3.2   Biological Networks Emerging from Computational Neuroscience: Sensory and Motor Modules

Computational neuroscientists have developed models of particular sensory or motor neural structures that could be incorporated into neural-based cognitive architectures. These models range from the early stages of auditory system (cochlea, brain stem, colliculus), to auditory thalamo-cortical models (Lyon and Shamma, 1996; Wang and Shamma, 1995; Shamma, 1997; Aleksandrovsky et al., 1996a). For vision, there are many visual system modules, including retina, geniculate, primary visual cortex and higher visual cortical areas (Landy and Movshon, 1991; Gupta and Knopf, 1994; Douglas et al., 1995). To enable visuo-motor behaviors, oculomotor system models (superior colliculus) have been developed linked to miniature cameras or retina-like vision sensors mounted in rapidly-slewing eye-balls (Schwartz et al., 1995). Saccadic eye movement generation has also been modeled (Ferrell, 1996; Gancarz and Grossberg, 1999). Some of these sensing structures have been emulated in analog or digital VLSI hardware for faster, parallel processing (Zornetzer et al., 1995). Motor system models are also emerging, from central pattern generators, to spinal loops, cerebellum, basal ganglia and motor cortex (Bullock et al., 1994, 1998; Contreras-Vidal et al., 1997; Alexander and Crutcher, 1990). These sensory and motor component systems should be evaluated as components of existing cognitive architecture (e.g. EPIC) designed to predict human sensori-motor performance. This would enable more sophisticated sensori-motor behaviors that adapt to context, but it is unlikely that existing cognitive architectures could exploit the power of these new sensor-motor models. I predict that a more powerful use of these emerging sensori-motor systems will be their use as building blocks of neural-based cognitive architectures. A cognitive architecture that incorporates such sensori-motor modules will enable these systems to function in a "real-world", not buffered by an enormous hand-crafted software interface. Such systems could operate directly on sensory data, take advantage of multimodal processing, temporal sequences, and scene level analysis. Moreover, they could implement true active sensing by completing sensory motor loop at all levels, rather than performing strictly bottom-up or top-down information processing. Active sensing is a key ingredient of "situated" cognition and autonomous cognitive systems (Brooks et al., 2000).

6.3.3   Forebrain Systems Supporting Cortical Function

Cognitive skills involve multiple cortical areas, and subcortical systems based both upon the neuropsychological literature and data emerging human brain imaging studies. These include: thalamus, basal ganglia, neuromodulator systems, limbic structures such as hippocampus and entorhinal cortex, the amygdala, and possibly the cerebellum. Many cortical models are actually thalamocortical models, some including ascending neuromodulator systems. There are a number of viable models of hippocampus and entorhinal cortex (Granger et al., 1989, 1996; Gluck and Granger, 1993; Gluck and Myers, 1997; Iatrou et al., 1999). There have also been a number of productive studies of role of neuromodulator systems (DA, NE), and recently there have been models of the basal ganglia developed that utilize the compelling anatomical and physiological data on the interaction of this system with the neocortex. Although models of these supporting neural structures are important in developing a system with cognitive capacity, based on the evolutionary expansion of neocortex in humans and the abundant neuropsychological evidence for cortical involvement, the development of a computationally effective model of general neocortex is the critical enabler for the creation of neural-based cognitive architectures.

6.4   Elements of a General Cortical Model

The construction of a general neocortical model with sufficient computational power
to execute cognitive skills, is likely to require the following elements:

1. processor elements that capture essential computational features
    of real neurons (McKenna et al., 1992; Koch, 1999);

2. microcircuitry based on specific connectivity motifs among multi-
    ple basic neuron types (excitatory and inhibitory);

3. dynamic synapses with adaptive filter properties;

4. ensemble coding and cortical dynamics to enable high capacity,
    yet agile, representations.

6.4.1   Single Neuron Models or Processor Elements

Real neurons are orders of magnitude more complex in their computational capabilities than the processing elements (PE) of artificial or connectionist neural networks.
The dendritic tree structure, active membrane biophysics, subcellular information
processing, enormous number of inputs and the flexibility of spike encoded outputs
delivered to large numbers of divergent synapses, each with a dynamics dependent on
the target neuron created imply that each neuron has extraordinary processing power.
The computations that can be performed by real neurons have been outlined elsewhere (McKenna et al., 1992; Koch, 1999; Segev et al., 1995). The challenge in cortical modeling is which features of neuronal geometry and biophysics to introduce
into the models. There are significant trade-offs in employing realistic neurons in network simulations. Complex neurons impact the speed of the network simulation and
the complexity of the simulation. On the other hand, realistic neurons are more adaptive, graded, are capable of multiprocessing, and they permit effective use of timing,
delay and multiplexed temporal patterns. In the case of biological networks for multi-
output process control and central pattern generators, the evidence is clear that maximal biophysical realism, complete with all known conductances and 2nd messenger
modulators, produces demonstrable advantages. The specification of how much of
the biophysics is needed for neocortical simulations is unknown, and is likely to
depend upon the selection of the microcircuitry and the synaptic dynamics.

6.4.2   Microcircuitry

The microcircuitry of primary visual cortex is well characterized, (Martin, 1988;
Lund et al., 1994; Braitenberg and Schuz, 1991) and a possible canonical microcircuit has been advanced (Martin, 1988; Douglas and Martin, 1990). The functional
interpretation of this microcircuitry is still a matter of active investigation even for

primitive functions like directional selectivity (Koch, 1999; Douglas et al., 1995; Segev et al., 1995).

Descriptions of the microcircuitry for the auditory cortex are emerging (Mitani et al., 1985; DeFelipe and Jones, 1988), however, for the frontal and parietal cortex there is less convergence on circuit details. The latter areas are difficult to address with single neuron physiology because they are not organized by simple primitives such as the sensory receptive fields in primary visual cortex. The simultaneous intracellular recording from multiple synaptically coupled neurons, combined with dye markers, provides a powerful technique for analyzing cortical microcircuitry, in terms of synaptic connections between neurons with specific morphologies and biophysics located within specific layers, including estimates of the number of active synapses between neurons as function of distance (Thompson and Deuchars, 1994; Braitenberg and Schuz, 1991). However, just as it seemed that a table of connectivity, and circuit motifs would emerge from these studies, the work foundered over the troubling fluctuations in synaptic efficacy observed in the experiments. This set the stage for the remarkable series of experiments by Markram and his associates demonstrating and characterizing the synaptic dynamics of cortico-cortical connections (see Chapter 5 by Markram, this volume).

6.4.3 Dynamic Synaptic Connectivity

The traditional view of cortical microcircuitry held that a small number (3–20) of different types of neurons, distinguished by morphological and neurochemical properties, were interconnected according to a set of motifs based on cortical layer (vertical) and intracortical distance (horizontal), and that these connections were basically stable in the adult, with some provision for stochastic synaptic transfer, and limited plasticity (such as LTP of synaptic efficacy). This view has been challenged recently by the seminal studies of cortical synaptic plasticity of Markram and his colleagues (see Chapter 5, this volume). The new picture that is emerging is that the plasticity of cortical synapses is not limited to changes in synaptic efficacy (strength), but that the synaptic transfer function is modified as a function of the precise timing of conjoint pre- and post- synaptic activity. This change in the basic filter properties of the synapse permit a high degree of multiplexing in networks and the rapid adaptive creation of small networks of neurons tuned to dynamic features of incoming activity patterns. This degree of synaptic dynamics could lead to more flexible and general extensions of binding by "synchrony" (Singer, 1994) or synfire chains (Abeles, 1991) in cortical neural networks. The computational capabilities of cortical networks employing these dynamic, nonlinear synapses is now being explored in large-scale simulations of spike-based cortical networks (Wills et al., 1999). The simple synaptology of cortical microcircuitry based on excitatory and inhibitory connections between specific cell types is now undergoing a major revision, and the rules governing synaptic transfer between specific cortical cells types is likely to be considerably more complex and dynamic than either the micro-anatomy or histochemistry had indicated.

6.4.4   Ensemble Dynamics and Coding

From the first decades of systematic study of electrically isolated single neurons a case emerged for the single neuron as the basic element of neural information coding, in line with Cajal's neuron doctrine. Analysis of the representation or coding of sensory and motor events was built from histograms of peri-event neuronal action potentials which were averaged over many trials (typically 20–30). In this view, multiunit (multiple neuron) recordings from lower impedance electrodes are viewed either as providing a larger sample of the population from nearby neurons, or were suspect when evidence was provided that nearby neurons might encode different ranges of the sensory-motor domain or even have totally different task dependencies. The latter findings raised the possibility that a complete microcircuit account of neuronal types, connectivities and functional activities might be needed to interpret the computational significance of the set of single neuron coding data. Recent electrophysiological technology that enable simultaneous recording from multiple neurons, coupled with analytical techniques to relate this activity to sensory stimuli or behavioral performance have produced increasing evidence that task relevant information is encoded in the distributed activity of neuronal ensembles. Evidence for distributed ensemble coding has been obtained in the entire somatosensory system, in inferotemporal and prefrontal cortex, and hippocampus (reviewed by Deadwyler and Hampson, 1997). Simultaneous multineuronal recordings have revealed that ensembles of neurons exhibiting a distributed representation of sensori-motor information may also exhibit epochs of synchronous oscillatory activity. The emerging data on population coding, and the role of synchronized activity gives important insight into the mesoscale operations of the cortex. Moreover, simultaneous population recording at various levels within a sensory system has provided strong evidence that sensory representations are not only distributed spatially, but temporally, and that recurrent connections mediate a consensus or settling of representations (see Nicolelis et al., 1998, 2001).

As brain imaging technology improves in spatial and temporal resolution, we may soon be able to directly compare neuronal ensemble activity and the mesoscale of brain activation visualized by imaging technologies in humans and non-human primates. Ensemble codes will likely prove crucial to interpreting mesoscale brain imaging data. Moreover, it may be possible to build cortical models based on neuronal ensembles as processing elements in lieu of the more slowly emerging cortical microcircuitry based on single neurons. Mosaic functional patterns of organization by activation of cortical minicolumns is another possible bridge to mesoscale brain imaging (Favorov et al., 1998). Not only is the phenomenon of microcolumn activation intriguing, but models based on microcolumn dynamics exhibit a number of desirable properties, including attractor-based sensory representations (Favorov et al., 1998).

6.4.5 Transient Coherent Structures and Cognitive Dynamics

Transient synchronization of oscillations or transient shifts in the pattern of cross-coherence between sites in cerebral cortex, including frontal and parietal sites, have been reported in conjunction with switches in sensori-motor or behavioral performance in cats, primates, and humans. I have previously reviewed some of this data and interpreted it within the context of the dynamics of nonlinear systems (McKenna et al., 1994). One point developed in that review worth reiterating is that complex systems, including neural systems, operating near instability have the ability to rapidly and reversibly shift global states in a controlled manner, and can exploit these properties for agile information processing, classification, pattern generation, coordination, and adaptive control. Moreover, synchronization of oscillating neural ensembles has been shown in connectionist models to permit dynamic cognitive binding (Shastri and Ajjanagadde, 1993).

The most compelling case for the importance of global neural dynamics in cognition comes from the four decade-long research program of Walter Freeman, exploring the nonlinear systems properties of the olfactory and limbic system. As Freeman reiterates in Chapter 3 of this book, neural information processing in paleocortical systems generates global attractor representations, and the chaotic brain dynamics observed in the olfactory system implies an active perception or expectation-based processing and evidence for internally generated goals. Moreover, the phase-space analysis of sensory representation in this system exhibits strong context-dependent processing. As the spatio-temporal resolution of brain imaging improves, the issues raised by Freeman on global neural dynamics will move to the foreground, and a complete account of the neural basis of cognitive skills will necessarily have to include characterization of the global and local attractor structure of cortical activity and its theoretical exegesis. It is heartening to note the level of interest in cognitive psychologists in dynamic systems, albeit largely at a heuristic level (Thelen and Smith, 1994; Port and van Gelder, 1995).

6.5 Promising Models and their Capabilities

6.5.1 Biologically Based Cortical Systems

There has been considerable effort focused on modeling the visual cortex, from modeling RF properties (e.g. orientation selectivity) of single biophysically realistic neurons (Douglas et al., 1995), to models that place the visual cortex into an entire functional model of the visual system, capable of high level visual processing of moving objects (Grossberg et al., 1997). The latter employ model neurons with limited biophysics, but placed in a network context that results in highly desirable computational primitives, from the viewpoint of machine vision. The range of visual phenomena and illusions that the neural models of Grossberg and his associates have accounted for is impressive. Visual modeling benefits in general from the enormous scale of vision neuroscience research, as well as the tangible sensory representations observed at the level of single neurons as well as entire cortical regions. It is less clear how to extend these visual cortical models into general cognitive architectures, since the desirable "feature" level neuronal representations are unknown for many higher cognitive processes, these models are constructed without reference to memory systems, and the computational basis of cognitive processes are likely to extend well beyond "representation" into generative, i.e., dynamic processes.

One of the more remarkable developments in computational neuroscience is the inordinate contribution of olfactory system models to the development of novel neural learning and pattern recognition algorithms, and successful demonstrations of the integration of hippocampal, limbic, paleocortical, and thalamocortical networks into functional computational systems that exploit the computational attributes of the component neural structures (Freeman, 1992, Chapter 3, this volume; White et al., 1998; White and Kauer, 1999; Pearce et al., 2001; Ambros-Ingerson et al., 1990).

One particularly promising series of cortical models has been developed by Granger and Lynch. The initial models grew out of simulations intended to understand the network consequences of LTP (long-term potentiation) rules emerging from hippocampal and olfactory cortex experiments. They first modeled a system consisting of an olfactory bulb connected to the olfactory paleocortex, with recurrence via the accessory olfactory nucleus (Ambros-Ingerson et al., 1990; Granger et al., 1989). The model included simple neurons, sparse connectivity, synapses with LTP, realistic time courses of PSPs, lateral inhibition produced winner-take-all (Coultrip et al., 1992), and recurrent inhibition, also via plastic synapses. Analysis of the pattern of activated cortical units, over successive cycles, showed that the responses were progressing down a classifier tree. In fact, as a classifier algorithm, the process matched or exceeded the efficiency, in terms of computational complexity, of the best known hierarchical clustering algorithms. Moreover, the model generated specific behavioral predictions, with positive test results (Granger et al., 1991a, 1991b).

They subsequently added modules for piriform/entorhinal cortex, and dentate, CA3 and CA1 of the hippocampus, each reflecting the known anatomical connectivity and specific type of synaptic plasticity for that structure. The resulting system exhibited the capability of performing hierarchical clustering on temporal input sequences (CA1 properties of temporal order learning), handling time-dilation (highly recurrent CA3 "holding memory") of signals, and exhibiting a very high-capacity for noise-tolerant data storage and retrieval of temporal sequences. The time-dilation property of CA3 also enabled autonomous segmentation of continuous speech by this system (Granger et al., 1996; Kilborn et al., 1998).

Building upon their experience with the paleocortical circuitry of the olfactory cortex, they built a neocortical model of the auditory cortex, combined with its supporting thalamic nuclei (Garzotto et al., 1997) (Figure 6.4). First, they identified the paleocortical circuitry embedded within the neocortex: the three-layer olfactory cortex was considered equivalent to the upper layers of auditory cortex. The upper layers receive non-topographic input to its layer 1 from the (plastic) magnocellular medial geniculate (MGm) nucleus, whereas the lower layers of auditory cortex receive strong inputs from the topographic (non-plastic) ventral medial geniculate (MGv) of the thalamus. Hence a model was constructed consisting of the specific (MGv) and non-specific (MGm) thalamus (the latter providing plastic NMDA synapses to cortex), thalamic reticular nucleus, and four layers of auditory cortex consisting of both excitatory and inhibitory neurons and a vertical columnar organization with synaptic connectivity within columns following the known microcircuitry, as well as cortico-cortical synaptic plasticity. Deep layers of cortex connect back to the MGm directly or via the inhibitory thalamic reticular nucleus to MGv. Due to the convergence of the plastic non-specific inputs and middle layers relaying specific thalamic inputs to the upper layers, these upper layers efficiently perform hierarchical clustering on signal ("cochlea") inputs, and then the cluster identities are passed to the deep layers, which construct a tree of cluster sequences. The deep layer also learns order dependent brief sequences corresponding to transitions between adjacent features. The outputs (signal sequences) learned in the primary auditory cortex can be relayed to secondary and tertiary auditory cortex modes, which then learn sequences of sequences, etc. Since the networks are sparsely connected, the entire signal is not represented continuously, but rather distinct segments are learned in distinct sub-networks. This auditory cortical system has been demonstrated to be a superior classifier of difficult continuous signals, such as clinical EEGs. So successful in fact, that algorithms are now regarded as proprietary. Hence, biologically realistic models of sensory cortex and associated structures are capable of demanding sensory processing and analysis. But how does one get to action and planning in such a system?

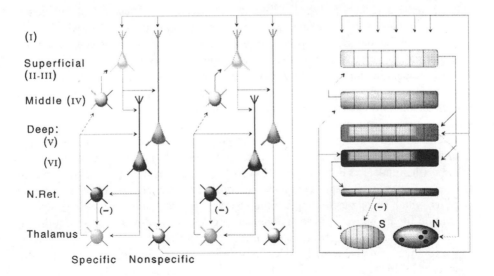

Figure 6.4. Components of a biologically derived model of the auditory thalamo-cortical system. Reproduced with permission from R. Granger, University of California at Irvine, 2002.

One path from sensing to acting is provided by adding a basal ganglia to the cortical and subcortical systems just described, and embedding the system in a mobile robot with vision. In fact, Granger and his associates have completed a very detailed model of a basal ganglia-cortical system consisting of matrix (matrisomal loop) and patch (striosomal loop) basal ganglia subsystems, and including links to substantia nigra DA inputs (reward and punishment), and motor outputs to control actions of the robot, as well as thalamo-cortical inputs (Granger, personal communication; Aleksandrovsky et al., 1997b). This system permits the mobile robot to learn to approach a hidden object, that is, navigate to a place where it can view the object.

6.5.2   A Cortical System Based on Neurobiology, Biological Principles and Mathematical Analysis: Cortronics

Hecht-Nielsen has developed a new theory of the cerebral cortex, embodied in a model named Cortronics (Hecht-Nielsen, 1998, Chapter 4, this volume). This model is presented in depth in this volume, but a concise account of its features is presented here because it simultaneously incorporates a great deal of neurobiology, the selection of the learning algorithms are well reasoned on mathematical grounds, and it has considerable potential for addressing cognitive skills.

This model has an elementary microcircuitry consisting of small pyramidal neurons in upper layers II–IV, large pyramidal neurons in lower layers IV–VI, and inhibitory interneurons. Broadly distributed layer I inputs arise from thalamic intralaminar nuclei which relay signals from the midbrain reticular formation, and specific thalamic inputs to the pyramidal neurons of the lower layers. An episodic process is postulated, termed restart, in which a barrage of ascending reticular input causes a competition for first-to-fire among the population of pyramidal neurons, which competitively inhibit neighboring pyramids via the interneurons. Since the pyramidal neurons are also receiving structured specific input from "sensory" systems via thalamus (or descending input from higher cortical areas), those clusters of neurons which win the competition come to form a representation of that input. Since the pyramids also have longer range recurrent connections, the winning clusters form a consensus spatial pattern of activation.

Hebbian-like learning rules operate to strengthen the synapses between co-active neurons, and the group of co-active neuron clusters then become "representational tokens" (the model has been shown to have a very large capacity of learned representations). Hierarchical representations emerge via connections to other cortical regions, which encode increasingly more abstract representations (chunking). Restart activation of the higher representations can provide a top-down expectation-driven processing, or activate a sequence of tokens. Moreover, many cortical regions converge onto a "frontal cortex" which develops "action" regions, which are then expressed as an "action token" for candidate actions. Action tokens compete for execution, with biases provided by drive-state-related inputs. The elaboration of action tokens into executed actions tailored to the current context takes place across a descending hierarchical cascade of regions, and at each region the tokens are elaborated into discrete temporal sequences of constitutive tokens, with specifics again influenced by current associative input.

A final competition then takes place in the motor system (pre-motor, motor cortex). The architecture can operate either in higher level pattern recognition of sensory sequences or as a generator of action (motor or "thought") sequences. Moreover, since the brain region which originates the restart signal, midbrain reticular formation, receives strong input from the cortex, there is an outer control loop (volition?) that can be utilized.

The model has been successfully demonstrated on context-dependent text sentence meaning representation, and machine vision (automatic segmentation, invariant classification, and "pop-out" recognition). This architecture has good prospects of being able to execute tasks requiring cognitive skills, and being able to bridge the gap between perception and action planning that has limited previous biological neural network models of cortex.

6.5.3 Connectionist Architectures with Biological Principles: The Convergence of Cognitive Science and Computational Neuroscience

While there have been a number of connectionist and hybrid neural network approaches developed for intelligent system applications or exploration of restricted cognitive domains, I have chosen to focus on one particular research stream that appears to be both exploiting biological principles and providing insight into the computational mechanisms of cognition. This is a research stream that begins with the collaborations between McClelland and Cohen on connectionist approaches to cognitive phenomena (e.g. Stroop effect) and later expands into models of active memory and controlled processing that combine distinct networks for hippocampus, prefrontal cortex, perceptuo-motor cortex and modulating systems (Cohen et al., 1990, 1996; Braver and Cohen, 1999, 2001; O'Reilly and Munakata, 2000).

Recently, O'Reilly (1998) explicitly laid out six principles for biologically based computational models of cortical cognition:

1. Biological realism: where possible computational models should be constrained and informed by biological properties of the cortex.

2. Distributed representations: Distributing representations over multiple neurons or processor elements provides the functional advantages that include efficiency, robustness, and accuracy, and the ability to represent similarity relationships.

3. Inhibitory competition: Mutual inhibition between neurons is important because it selects representations for processing, enables subsequent refinement of the representations during learning, and leads to sparse representations.

4.   Bidirectional activation propagation (recurrence): The communication of activation simultaneously in both bottom-up and top-down directions.

5.   Error-driven task learning: This is important for shaping representations according to task demands by learning to minimize the difference between a desired outcome and the network output. (He gets around the usual critique of such algorithms being non-biological by demonstrating a biologically plausible variant of the algorithm that exploits the bidirectional activation principle).

6.   Hebbian model learning: This serves to form internal representations of the general structure of the environment, without respect to particular tasks (it should be noted that these principles parallel those put forward by Hecht-Nielsen in his Cortronic model, discussed above).

O'Reilly has also developed a network model which implements all these principles as an algorithm named Leabra (O'Reilly, 1998; O'Reilly and Munakata, 2000). He has also demonstrated that three of these principles, namely, inhibitory competition, Hebbian learning, and bidirectional connectivity, when combined have significant effects on network generalization on combinatorial problems (O'Reilly, 2001) and also reading tasks (O'Reilly, personal communication).

In collaboration with Cohen and Braver, these neural principles have been incorporated into models of frontal cortex, communicating with hippocampus and neuromodulator systems (O'Reilly et al., 1999). The specific hypothesis is that the prefrontal cortex, together with other essential neural systems, plays a critical role in cognitive control. Cognitive control is taken to mean the ability of the cognitive system to flexibly adapt its behavior to the demands of specific tasks, including directing the processing of task-relevant information and the mediation of task-relevant behavior (O'Reilly et al., 1999; O'Reilly and Rudy, 2001; O'Reilly et al., 2002).

This controlled processing is treated as emerging from the interactions of several brain systems. One such component is the posterior perceptual and motor cortex (PMC). The PMC optimizes knowledge-dependent inference capabilities, which depend on dense interconnectivity, highly distributed representations, and slow integrative learning to obtain good representations of important statistical properties of the world. Sensori-motor and multimodal processing proceeds in a hierarchical but highly interconnected fashion. The prefrontal cortex (PFC) component is specialized for active maintenance of internal contextual information that is dynamically updated and self regulated. The PFC optimizes active memory via restricted recurrent excitatory connectivity and a gating mechanism (initially the dopaminergic neuromodulating system, and more recently the basal ganglia). This confers the ability to flexibly update internal representations, to maintain these over time in the face of interference and, by propagation of activation, to bias PMC processing in a task dependent manner. The representations in PFC are relatively isolated from one another and can be activated in a combinatorial manner. Only a small number of representations are actively maintained at any given moment and inhibitory "attentional" mechanisms are posited. More recent versions of the PFC model exploit activation-based processing based on matching features to inputs versus weight-based processing.

Another important component is the hippocampus and related structures (HCMP). The HCMP model optimizes rapid learning of arbitrary information in weight-based memories, permitting the binding of elements of a novel association, including representations in PFC and PMC, providing a mechanism for temporary storage of arbitrary current states for later retrieval. Learning in the HCMP uses pattern separation, with separation of individual learning episodes, in contrast to the integrative nature of the PFC. This pattern separation requires sparse, conjunctive representations with all the elements contributing to the representation. This conjunctivity leads to context-specific and episodic memories, which bind together the elements of a context or episode. This is in contrast to the combinatorial PFC in which the elements contribute in a separable manner.

From this system of interacting forebrain components, the controlled processing of task-relevant information emerges as: (1) the use of and updating of actively maintained representations in the PFC biasing the subsequent processing and activation within PMC in a task-appropriate manner, and (2) the interplay of the PFC biasing and HCMP binding (O'Reilly et al., 1999). This forebrain system model has been successfully applied to dynamic categorization tasks (based on the Wisconsin card sorting task) in which the rules governing a target sequence are changed without warning and the subject or model must update the categorization rule. This system, particularly with the addition of the basal ganglia component to provide detailed control of prefrontal cortex, shows great promise in being able to eventually achieve analogical reasoning.

6.6   The Challenges of Demonstrating Cognitive Ability

Given that one can develop a neural-based cognitive architecture, how would one demonstrate that it could execute cognitive skills and what metric would one apply? Again, the connectionists have provided some significant techniques. One can simply take a task designed to test human cognitive skills, and modify it to enable the model to provide the solutions. The danger is that one can seem to be addressing toy problems in an artificial world. How does one know that some level of general cognitive skill has been achieved? There exist tests that seem to tap general central skills that are amenable to presentation to non-literate computer simulations.

Snow et al., (1984) performed a non-metric scaling analysis of a large number of tests of human ability; those tests at the center of a two-dimensional scaling solution test mental abilities in a manner less biased by test format, and these also tended to be complex reasoning tasks. The most central of the ability tests was the Raven's Progressive Matrices Test, which utilizes analogical reasoning for spatial pattern completion of arrays of complex visual icons. Such a test should be tractable for presentation to machines or non-human primates. Carpenter and Just have analyzed the Raven's test performance and concluded that individual differences in problem solving are largely accounted for by differences in working memory capacity (Carpenter et al., 1990).

While tests such as Raven's Progressive Matrices could provide an informative challenge to neural-based cognitive architectures, and serve as a performance benchmark, the ultimate challenge for autonomous intelligent systems will be autonomous mobile robots.

6.6.1   Robotics and Autonomous Systems

The tasks expected of autonomous mobile system have much in common with animal intelligence (e.g. navigation, landmark and route learning, threat detection and assessment, foraging strategies, hunting, hiding, cooperative behavior, deception). Moreover, active sensing is emerging as a dominant theme in mobile platform based surveillance, and active sensing in animals has been studied by neuroethological studies of neural specializations such as biosonar and oculomotor control. Hence, neural-based cognitive architectures based on the analysis of animal nervous systems are particularly good candidates for the inspiration and development of intelligent systems for mobile robots. For many applications, such mobile robots may require a different set of cognitive skills more akin to animal intelligence in addition to human cognitive skills.

A truly autonomous cognitive architecture requires the combination of active sensing, motor control, and task-dependent reconfigurable cognitive processors to enable it to function in, and operate upon, the real real-world.

6.7   Co-development Strategies for Automated Systems and Human Performers

Ultimately, I foresee the development of neurobiologically based cognitive architectures implemented for efficient, adaptive modeling of human abilities. These cognitive architectures would consist of ultra-scale neocortex emulating networks implemented in silicon, capable of executing a wide range of human-like cognitive skills. These architectures would have a dual role: (1) they would serve as models of human performers for design of human-centered systems (such as performance and decision aids and embedded trainers) and computational engines within intelligent tutors; (2) they would serve as the reasoning engine of powerful intelligent systems, sufficiently compact so as to be implemented on autonomous robots. These cognitive architectures would be combined with artificial sensing systems (hyperspectral vision, audition, vestibular, RF) so that they would be fully capable of processing raw sensor data, and controlling the platforms and sensors, and hence be capable of autonomous navigation, situation assessment, and planning. This dual use of a common neural-based cognitive architecture would enable direct trade-offs of cognitive task delegation between human users and intelligent systems, enabling the best use of human capabilities and also enhancing overall system performance.

6.8 Acknowledgements

I wish to thank Dr Susan Chipman for her helpful comments.

References

Abeles, M. (1991) Corticonics. Cambridge, England: Cambridge University Press.

Abeles, M., Vaadia, E., Bergman, H., Prut, Y., Haalman, I., Slovin, H. (1993) Dynamics of neuronal interactions in the frontal cortex of behaving monkeys. Concepts in Neuroscience 4: 131–158.

Aleksandrovsky, B., Whitson, J., Andes, G., Lynch, G., Granger, R. (1996a) Novel speech processing mechanism derived from auditory neocortical circuit analysis. Proc. ICSLP Int'l. Conf. On Spoken Language Proc. IEEE Press 1: 558–561.

Aleksandrovsky, B., Whitson, J., Garzotto, A., Lynch, G., Granger, R. (1996b) An algorithm derived from thalamocortical circuitry stores and retrieves temporal sequences. Proc. Int'l. Conf. Pattern Recog., IEEE Comp. Soc. Press 4: 550–554.

Aleksandrovsky, B., Brucher, F., Lynch, G. (1997a) Neural network model of striatal complex. In: Biological and Artificial Computation: From Neuroscience to Technology. IWANN'97 International Conference on Artificial and Natural Neural Networks, Lecture Notes in Computer Science 1240. Berlin: Springer-Verlag, pp. 103–115.

Aleksandrovsky, B., Whitson, J., Garzotto, A., Lynch, G., Granger, R. (1997b) A continuous temporal sequence recognition device based on a model of structure and function in the neocortex. Technical Report, Brain Theory Project, University of California, Irvine.

Alexander, G.E., Crutcher, M.D. (1990) Functional architecture of basal ganglia circuits: Neural substrates of parallel processing. Trends in Neurosciences 13: 266–271.

Ambros-Ingerson, J., Granger, R., Lynch, G. (1990) Simulation of paleocortex performs hierarchical clustering. Science 247: 1344–1348.

Anderson, J.R. (1993) Rules of the Mind. Hillsdale, NJ: Erlbaum.

Anderson, J.R., Matessa, M., Labiere, C. (1997) ACT-R: A theory of higher level cognition and its relation to visual attention. Human Computer Interaction 12: 439–462.

Angeline, P., Saunders, G., Pollack, J. (1994) An evolutionary algorithm that constructs recurrent networks. IEEE Trans. on Neural Networks 5: 54–65.

Anton, P., Lynch, G., Granger, R. (1991) Computation of frequency-to-spatial transform by olfactory bulb glomeruli. Biol. Cybern. 65: 407–414.

Bachmann, C.M., Musman, S., Luong, D., Schultz, A. (1994) Unsupervised BCM projection pursuit algorithms for classification of simulated radar presentations. Neural Networks 7: 709–728.

Bowers, J.M., Beeman, D. (Eds.) (1998) The Book of Genesis: Exploring Realistic Neural Models with the General Neural Simulation System. 2nd ed. New York: Springer-Verlag.

Braitenberg, V., Schuz, A. (1991) Anatomy of the Cortex: Statistics and Geometry. Berlin: Springer-Verlag.

Braver, T.S., Cohen, J.D. (1999) On the control of control: The role of dopamine in regulating prefrontal function and working memory. In: S. Monsell, J. Driver (Eds.) Attention and Performance XVII. Cambridge, MA: MIT Press.

Braver, T.S., Cohen, J.D. (2001) Working memory, cognitive control, and the prefrontal cortex: computational and empirical studies. Cognitive Processing 2: 25–55.

Brooks, R., Breazeal, C., Marjanović, M., Scassellati, B., Williamson, M. (1999) The Cog project: Building a humanoid robot. In: C. Nehaniv (Ed.) Computation for Metaphors, Analogy, and Agents, Lecture Notes in Artificial Intelligence 1562. New York: Springer-Verlag, pp. 52–87.

Bullock, D., Cisek, P.E., Grossberg, S. (1998) Cortical networks for control of voluntary arm movements under variable force conditions. Cerebral Cortex 8: 48–62.

Bullock, D., Fiala, J.C., Grossberg, S. (1994) A neural model of timed response learning in the cerebellum. Neural Networks 7: 1101–1114.

Carpenter, G.A., Grossberg, S. (Eds.) (1991) Pattern Recognition by Self-Organizing Neural Networks. Cambridge, MA: The MIT Press.

Carpenter, P.A., Just, M.A., Shell, P. (1990) What one intelligence test measures: A theoretical account of the Raven Progressive Matrices Test. Psychological Review 97: 404–431.

Chandrasekaran, A. (1994) Architecture of Intelligence: The problems and current approaches to solutions. In: V. Honavar, L. Uhr (Eds.) Artificial Intelligence and Neural Networks. New York: Academic Press.

Chipman, S.F. (1992) The higher-order cognitive skills: What they are and how they might be transmitted. In: T.G. Sticht, B.A. McDonald, M.J. Beeler (Eds.) Intergenerational Transfer of Cognitive Skills: Vol. II: Theory and Research in Cognitive Science. Norwood, NJ: Ablex, pp. 128–158.

Cohen, J.D., Braver, T.S., O'Reilly, R.C. (1996) A computational approach to prefrontal cortex, cognitive control and schizophrenia: recent developments and current challenges. Phil. Trans. R. Soc. Lond. B. 351: 1515–1527.

Cohen, J.D., Dunbar, K., McClelland, J.L. (1990) On the control of automatic processes: A parallel distributed processing model of the stroop effect. Psychological Review 97: 332–361.

Contreras-Vidal, J.L., Grossberg, S., Bullock, D. (1997) A neural model of cerebellar learning for arm movement control: Cortico-spinal-cerebellar dynamics. Learning and Memory 3: 475–502.

Cooper, R., Shallice, T. (1995) Soar and the case for unified theories of cognition. Cognition 55: 115–149.

Coultrip, R., Granger, R., Lynch, G. (1992) A cortical model of winner-take-all competition via lateral inhibition. Neural Networks 5: 47–54.

Cox, A.L. and Young, R.M. (2000) Device-oriented and task oriented exploratory learning of interface designs. In: Proceedings of the Third International Conference on Cognitive Modeling. Veenendaal, Netherlands: Universal Press, pp. 70–77.

Deadwyler, S.A., Hampson, R.E. (1997) The significance of neural ensemble codes during behavior and cognition. Ann. Rev. Neurosci. 20: 217–244.

DeFelipe, J., Jones, E.G. (1988) Cajal on the Cerebral Cortex: An annotated translation of the complete writings. New York: Oxford University Press.

Dehaene, S., Changeux, J.P. (1992) The Wisconsin card sorting test: Theoretical analysis and modeling in a neuronal network. Cerebral Cortex 1: 62–79.

Douglas, R.J., Koch, C., Mahowald, M., Martin, K., Suarez, H. (1995) Recurrent excitation in neocortical circuits. Science 269: 981–985.

Douglas, R.J., Martin, K.A.C. (1990) Neocortex. In: G.M. Sheperd (Ed.) The Synaptic Organization of the Brain, 3rd Ed. New York: Oxford University Press, pp. 389–438.

Favorov, O.V., Hester, J.T., Kelly, D.G., Tommerdahl, M., Whitsel, B.L. (1998) Lateral interactions in cortical networks. In: M.J. Rowe (Ed.) Somatosensory Processing: From Single Neuron to Brain Imaging. Langhorne, PA: Harwood, pp. 187–207.

Feldman, J.A. (1991) Cognition as search. Science 251: 575.

Ferrell, C. (1996) Orientation behavior using registered topographic maps. In: From Animals to Animats, Proc. 1996 meeting of Soc. of Adaptive Behavior. Cape Cod, MA, pp. 94–103.

Frank, M., Loughry, B., O'Reilly, R.C. (2001) Interactions between the frontal cortex and basal ganglia in working memory: A computational model. Cognitive, Affective, and Behavioral Neuroscience 1: 137–160.

Freeman, W.J. (1992) Tutorial in neurobiology: From single neurons to brain chaos. Int. J. Bifurcation and Chaos 2: 451–482.

Gancarz, G., Grossberg, S. (1999) A neural model of saccadic eye movement control explains task-specific adaptation. Vision Res. 39: 3123–3143.

Garzotto, A., Aleksandrovsky, B., Lynch, G., Granger, R. (1997) A neocortically derived model of continuous contextual processing. Proc. International Conference on Neural Networks, IEEE Press 1: 564–568.

Gluck, K.A. (2000) An ACT-R/PM model of algebra symbolization. In: N. Taatgen, J. Aasman (Eds.) Proceedings of the Third International Conference on Cognitive Modeling. Veenendaal, Netherlands: Universal Press, pp. 134–141.

Gluck, M.A., Granger, R. (1993) Computational aspects of the neural bases of learning and memory. Annual Review of Neuroscience 16: 667–706.

Gluck, M.A., Myers, C.E. (1997) Psychological models of hippocampal function in learning and memory. Annual Review of Psychology 48: 481–514.

Granger, R., Ambros-Ingerson, J., Lynch, G. (1989) Derivation of encoding characteristics of layer II cerebral cortex. J. Cognit. Neurosci. 1: 61–87.

Granger, R., Cobas, A., Lynch, G. (1991a) Possible computations of primary sensory cortex: Hypotheses based on computer models of olfaction and audition. In: M. Baudry, J. Davis (Eds.) Current Issues in LTP. Cambridge, MA: MIT Press.

Granger, R., Staubil, U., Powers, H., Otto, T., Ambros, J., Lynch, G. (1991b) Behavioral tests of a prediction from a cortical network simulation. Psychological Science 2: 116–118.

Granger, R., Wiebe, S.P., Taketani, Lynch, G. (1996) Distinct memory circuits composing the hippocampal region. Hippocampus 6: 567–578.

Grossberg, S. (1997) Cortical dynamics of three-dimensional figure-ground perception of two-dimensional pictures. Psychol. Rev. 104: 618–658.

Grossberg, S., Mingolla, E., Ross, W.D. (1997) Visual brain and visual perception: how does the cortex do perceptual grouping? Trends in Neurosciences 2: 106–111.

Gunzelmann, G., Anderson, J.R. (2001) Modeling the emergence of strategies and their effects on problem difficulty in ACT-R. In: Proceedings of the Fourth International Conference on Cognitive Modeling. Mahwah, NJ: Lawrence Erlbaum Associates, pp.109–114.

Gupta, M.M., Knopf, G.K. (1994) Neuro-Vision Systems: Principles and Applications. New York: IEEE Press.

Hecht-Nielsen, R. (1998) A theory of the cerebral cortex. Proc. 1998 Int'l. Conf. on Neural Information Processing. Kitakyushu, Japan, pp. 1459–1464.

Iatrou, M., Berger, T.W., Marmarelis, V.Z. (1999) Application of novel modeling method to the nonstationary properties of potentiation in the rabbit hippocampus. Annals of Biomedical Engineering 27: 581–591.

Just, M.A., Carpenter, P.A. (1985) Cognitive coordinate systems: Accounts of mental rotation and individual differences in spatial ability. Psychological Review 92: 137–172.

Just, M.A., Carpenter, P.A. (1992) A capacity theory of comprehension: Individual differences in working memory. Psychological Review 99: 122–149.

Just, M.A., Carpenter, P.A., Hemphill, D.D. (1996) Constraints on processing capacity: Architectural or implementational? In: D. Steier, T. Mitchell (Eds.) Mind Matters: A Tribute to Allan Newell. Mahwah, NJ: Erlbaum.

Just, M.A., Carpenter, P.A., Shell, P. (1990) What one intelligence test measures: A theoretical account of the processing in the Raven Progressive Matrices Test. Psychological Review 97: 404–431.

Just, M.A., Carpenter, P.A., Varma, S. (1999) Computational modeling of high-level cognition and brain function. Human Brain Mapping 8: 128–136.

Kilborn, K., Kubota, D., Lynch, G., Granger, R. (1998) Parameters of LTP induction modulate network categorization behavior. In: J.M. Bowers (Ed.) Computational Neuroscience: Trends in Research. New York: Plenum Press, pp. 353–358.

Koch, C. (1999) Biophysics of Computation: Information Processing in Single Neurons. New York: Oxford University Press.

Labiere, C. (1998) The dynamics of cognition: An ACT-R model of cognitive arithmetic. PhD Dissertation. CMU Computer Science Dept. Technical Report CMU-CS-98-186. Pittsburgh, PA. (http://reports-archive.adm.cs.cmu.edu/cs1998.html)

Laird, J.E., Newell, A., Rosenbloom, P.S. (1987) SOAR: An architecture for general intelligence. Artificial Intelligence 33(1): 1–63.

Landy, M.S., Movshon, J.A. (1991) Computational Models of Visual Processing. Cambridge, MA: MIT Press.

Lund, J.S., Yoshioka, T., Levitt, J.B. (1994) Substrates for interlaminar connections in area V1 of macaque monkey cerebral cortex. In: A.A. Peters, K.S. Rockland (Eds.) Cerebral Cortex v. 10. New York: Plenum Press.

Lyon, R., Shamma, S. (1996) Auditory representations of timbre and pitch. In: H.L. Hawkins, T.A. McMullen, A.N. Popper, R.R. Fay (Eds.) Auditory Computation. Berlin: Springer-Verlag, pp. 221–270.

Martin, K.A.C. (1988) The Wellcome Prize lecture: from single cells to simple circuits in the cerebral cortex. Quart. J. Exp. Physiol. 73: 637–702.

McKenna, T.M. (1994) The role of interdisciplinary research involving neuroscience in the development of intelligent systems. In: V. Honavar, L. Uhr (Eds.) Artificial Intelligence and Neural Networks. New York: Academic Press.

McKenna, T., Davis, J., Zornetzer, S.F. (1992) Single Neuron Computation. Boston: Academic Press.

McKenna, T.M., McMullen, T.A., Shlesinger, M.F. (1994) The brain as a dynamic physical system. Neuroscience 60: 587–605.

Meyer, D.E., Kieras, D.E. (1997) A computational theory of executive cognitive processes and human multiple-task performance: Part 1. Basic mechanisms. Psychological Review 104: 3–65.

Mitani, A., Shimokouchi, M., Itoh, K., Nomura, S., Kudo, M. Mizuno, N. (1985) Morphology and laminar organization of electrophysiologically identified neurons in the primary auditory cortex. J. Comp. Neurol. 235: 430–447.

Newell, A. (1992) Unified theories of cognition and the role of Soar. In: J.A. Michon, A. Akyürek (Eds.) SOAR: A Cognitive Architecture in Perspective. Dordrecht, The Netherlands: Kluwer Academic Publishers, pp. 25–79.

Nicolelis, M.A.L., Baccala, L.A., Lin, R.C.S., Chapin, J.K. (1995) Sensorimotor encoding by synchronous neural ensemble activity at multiple levels of the somatosensory system. Science 268: 1353–1358.

Nicolelis, M.A.L., Fanselow, E.E., Ghazanfar, A.A. (1997) Hebb's dream: The resurgence of cell assemblies. Neuron 19: 219–221.

Nicolelis, M.A.L., Fanselow, E.E., Shuler, M. Henriquez, C. (2001) A critique of the pure feedforward model of touch. In: R.J. Nelson (Ed.) The Somatosensory System: Deciphering the Brain's Own Body System. Boca Raton, FL: CRC Press.

Nicolelis, M.A.L., Katz, D., Krupa, D.J. (1998) Potential circuit mechanisms underlying concurrent thalamic and cortical plasticity. Rev. Neurosci. 9: 213–224.

O'Reilly, R.C. (1996) Biologically plausible error-driven learning using local activation differences: The generalized recirculation algorithm. Neural Comput. 8: 895–938.

O'Reilly, R.C. (1998) Six principles for biologically based computational models of cortical cognition. Trends in Cognitive Sciences 2: 455–462.

O'Reilly, R.C. (2001) Generalization in interactive networks: The benefits of inhibitory competition and Hebbian learning. Neural Computation 13: 1199–1241.

O'Reilly, R.C., Braver, T.S., Cohen, J.D. (1999). A biologically-based computational model of working memory. In: A. Miyake, P. Shah (Eds.) Models of Working Memory: Mechanisms of Active Maintenance and Executive Control. New York: Cambridge University Press, pp. 375–411.

O'Reilly, R.C., Munakata, Y. (2000) Computational Explorations in Cognitive Neuroscience: Understanding of the Mind by Simulating the Brain. Cambridge, MA: MIT Press.

O'Reilly, R.C., Noelle, D.C., Braver, T.S., Cohen, J.D. (2002) Prefrontal cortex in dynamic categorization tasks: Representational organization and neuromodulatory control. Cerebral Cortex 12: 246-257.

O'Reilly, R.C., Rudy, J.W. (2001) Conjunctive representations in learning and memory: Principles of cortical and hippocampal function. Psychological Review 108: 311–345.

Pearce, T.C, Vershure, P.F.M.J., White, J., Kauer, J.S. (2001) Stimulus encoding during the early stages of olfactory processing: A modeling study using an artificial olfactory system. Neurocomputing 38: 299–306.

Port, R.F., van Gelder, T. (1995) Mind as Motion: Explorations in the Dynamics of Cognition. Cambridge, MA: MIT Press.

Rosenbloom, P.S., Laird, J.E., Newell, A. (1993) The Soar Papers: Readings on Integrated Intelligence. Cambridge, MA: MIT Press.

Salvucci, D.D., Boer, E.R., Liu, A. (2002) Toward an integrated model of driver behavior in a cognitive architecture. Transportation Research Record (in press).

Scassellati, B. (1998) A binocular, foveated active vision system. Technical Report, Memo 1628, MIT Artificial Intelligence Lab.

Schoenbaum, G., Eichenbaum, H. (1995) Information coding in the rodent prefrontal cortex. II. Ensemble activity in the orbitofrontal cortex. J. Neurophysiol. 70: 28–36.

Schunn, C., Anderson, J.R. (1998) Scientific discovery. In: J.R. Anderson, C. Labiere (Eds.) The Atomic Components of Thought. Mahwah, NJ: Erlbaum, pp. 255–296.

Schunn, C., Harrison, A. (2001) ACT-RS: A neuropsychologically inspired model for spatial reasoning. In: Proceedings of the Fourth International Conference on Cognitive Modeling. Mahwah, NJ: Lawrence Erlbaum Associates, pp. 267–268.

Schwartz, E.L., Greve, D.N., Bonmassar, G. (1995) Space-variant active vision: Definition, overview and examples. Neural Networks 8: 1297–1308.

Segev, I., Rinzel, J., Sheperd, G. (1995) The Theoretical Foundation of Dendritic Function: Selected Papers of Wilfred Rall with Commentaries. Cambridge, MA: MIT Press.

Shamma, S. (1997) Auditory cortical representation of complex acoustic spectra as inferred from the ripple analysis method. Network: Computation in Neural Systems 7: 439–476.

Shastri, L., Ajjanagadde, V. (1993) From simple associations to systematic reasoning: a connectionist representation of rules, variable, and dynamic bindings using temporal synchrony. Behav. Brain Sci. 16: 417–494.

Singer, W. (1994) Putative functions of temporal correlations in neocortical processing. In: C. Koch, J. Davis (Eds.) Large-Scale Neuronal Theories of the Brain. Cambridge, MA: MIT Press, pp. 201–237.

Snow, R.E., Kyllonen, P.C., Marshalek, B. (1984) The topography of ability and learning correlations. In: R.J. Sternberg (Ed.) Advances in the Psychology of Human Intelligence, Vol. 2. Hillsdale, NJ: Erlbaum, pp. 47–103.

Sohn, M.-H., Ursu, S., Anderson, J.R., Stenger, V.A., Carter, C.S. (2000) The role of prefrontal cortex and posterior parietal cortex in task-switching. Proceedings of National Academy of Science, 13448–13453.

Thelen, E., Smith, L.B. (1994) A Dynamic Systems Approach to the Development of Cognition and Action. Cambridge, MA: MIT Press.

Thibadeau, R., Just, M.A., Carpenter, P.A. (1982) A model of the time course and content of reading. Cognitive Science 6: 157–203.

Thomson, A.M., Deuchars, J. (1994) Temporal and spatial properties of local circuits in neocortex. Trends in Neurosciences 17: 119–126.

Vincente, K.J., Kirklik, A. (1992) On putting the cart before the horse: Taking perception seriously in unified theories of cognition. Behavioral and Brain Sciences 15: 461–462.

Wang, K., Shamma, S. (1995) Representation of acoustic signals in the primary auditory cortex. IEEE Trans. Audio and Speech Proc. V3(5): 382–395.

Weinberger, N.M. (1995) Dynamic regulation of receptive fields and maps in the adult sensory cortex. Ann. Rev. Neurosci. 18: 129–158.

Weinberger, N.M., Ashe, J.H., Metherate, R., McKenna, T.M., Diamond, D.M., Bakin, J.S., Lennartz, R.C., Cassady, J.M. (1990) Neural adaptive information processing: A preliminary model of receptive-field plasticity in auditory cortex during Pavlovian conditioning. In: M. Gabriel, J. Moore (Eds.) Learning and Computational Neuroscience: Foundations of Adaptive Networks. Cambridge, MA: MIT Press, pp. 91–138.

White, J., Dickinson, T.A., Walt, D.R., Kauer, J.S. (1998) An olfactory neuronal network for vapor recognition in an artificial nose. Biological Cybernetics 78: 245–251.

White, J., Kauer, J.S. (1999) Odor recognition in an artificial nose by spatio-temporal processing using an olfactory neuronal network. Neurocomputing 26: 919–924.

Wickens, J. (1997) Basal ganglia: Structure and computations. Network: Computation in Neural Systems 8: 77–109.

Wills, H.R., Kellogg, M.M., Goodman, P.H. (1999) A biologically realistic computer of neocortical associative learning for the study of aging and dementia. J. Invest. Med. 47(2): 11A

Zachary, W., Ryder, J., Hicinbothom, J. (1998) Cognitive task analysis and modeling of decision making in complex environments. In: J. Cannon-Bowers, E. Salas (Eds.) Decision Making Under Stress: Implications for Training and Simulation. Washington, DC: American Psychological Association Press.

Zornetzer, S., Davis, J., Lau, C., McKenna, T. (Eds.) (1995) An Introduction to Neural and Electronic Networks (2nd Ed.). San Diego: Academic Press.

Chapter 7

The Behaving Human Neocortex as a Dynamic Network of Networks

Jeffrey P. Sutton and Gary Strangman

7.1 Abstract

The neocortex is arguably the most sophisticated structure within the mammalian brain. It is the largest brain structure in the human, and its properties endow us with qualities that are unique to our species. In order to develop a systems-level approach to understanding the neocortex, and to develop a theoretical basis that is both tractable and useful, enormous anatomical and physiological simplifications must be made. These include focusing on specific aspects of function, such as associative memory and learning, at the expense of many other important characteristics. Such simplifications are required in any systems-level model of the brain.

The main contribution of this chapter is to describe a model of the neocortex that links together different scales or levels of neural organization. Specifically, individual neurons cluster together into networks, and these networks cluster together into larger networks, and so on. We argue that dynamically reconfigurable networks of neurons exist at multiple scales, and are fundamental to the structural and functional integrity of the cortex.

Experimental data supporting both the approach and the predictions of our neocortical model, termed the Network of Networks (NoN), are described. Section 7.2 summarizes some of the neurobiology relevant to the theoretical approach. The NoN is described in Section 7.3, where it is suggested that the model should do more than simply provide descriptors of neocortical organization function. Instead, it should lead to new ways of understanding and applying rapid, parallel and associative computations within and between neural networks at different scales. This is discussed in Section 7.4, where we investigate the model's veracity in tests of predicatability and falsifiability. In the final section, implications of the NoN for neuroengineering are considered.

7.2   Neural Organization Across Scales

The human brain consists of approximately 10^{11} neurons, with highly complex patterns of interconnections. A preponderance of these cells and connections are in the neocortex. Approximately 99 percent of the connections between cells in the cortex arise from other cells within the cortex, leaving only 1 percent of the connections originating from or terminating onto non-cortical structures (Shepherd, 1990). Thus, the neocortex is largely a self-contained structure.

The neurons within the neocortex display remarkable heterogeneity (Connors and Gutnick, 1990; Gupta et al., 2000). Some cells have relatively few connections confined to distances of less than 100 microns, while other cells, such as the excitatory pyramidal cells, have more than 10^5 connections located throughout arborizations extending from several centimeters to more than a meter (Brodal, 1981). Connectivity among cells forms networks that have complex anatomical and physiological properties at multiple scales of organization. Feedforward and feedback loops within and between networks are common throughout the neocortex. There are also critical loops between the neocortex and other structures, such as the thalamus and basal ganglia. Oscillations in these loops encode and manifest important timing mechanisms.

Within the neocortex, there is a degree of translational invariance in structure. Almost all areas have six layers of cells arranged parallel to the cortical surface and, with the exception of primary visual cortex, the general cellular composition of the neocortex is remarkably consistent across its surface, and even across species (Rockel et al., 1980). Despite this general uniformity, it has been known for almost a century that different regions of the neocortex have cytoarchitectural differences, as shown in Figure 7.1A (Brodmann, 1909). Functional correlates of these regions have been identified, and in several instances, functional networks have been mapped. This mapping provides one level of organization intermediate between single neurons and the entire network.

A variety of anatomical and functional properties have served to identify additional levels of organization within and across these Brodmann areas. For example, within such regions one can identify clusters of neurons, called cortical columns (illustrated in Figure 7.1B), which can be treated as anatomical, functional or computational units (Szentagothai, 1977). More than thirty regions are known to be associated with visual processing (Van Essen et al., 1992). Networks have also been identified and characterized based, in part, on selective cellular responses to small ranges of particular stimuli (e.g., Ts'o et al., 1990; Georgopoulos et al., 1993; Tanaka, 1993; Nicolelis et al., 1998; Pouget et al., 2000; Sanes and Donoghue, 2000). Small populations of neurons functionally cluster together in space and/or time to encode complex stimuli, and to predict, plan and carry out responses. Dynamic patterning has been found to be important for the representation of odors (Friedrich and Laurent,

2001), and dynamical features have been described for visual evoked responses (Makeig et al., 2002), somatosensory responses (Nicolelis et al., 1998; Chapin and Nicolelis, 1999), and in working memory (Cohen et al., 1997; Courtney et al., 1997). Moreover, there is ongoing plasticity among cells involving a variety of processes that include, but are not limited to, long-term potentiation and long-term depression (Brown et al., 1990; Kandel et al., 1991).

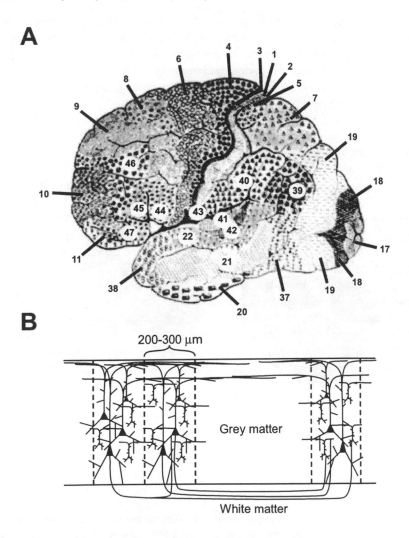

Figure 7.1. (A) The neocortex is subdivided into regions based on microscopic differences in the ways cells are arranged. A commonly used map is the one developed by Brodmann (1909), shown here, which – despite the general uniformity of cortex – identifies over 40 separate sub-regions, providing one intermediate level of spatial scale. (B) Within such regions, the neocortex shows further anatomical organization – a repeating grid of modules appears with dimensions on the order of 200–300 μm, known as cortical columns. Neurons within a column tend to be more tightly connected than between columns, as illustrated.

Cortical responses involve dynamic circuitry, with a high degree of parallel processing and interaction across scales. New neuroimaging technologies, such as functional magnetic resonance imaging (fMRI), have made it possible to look for large scale functional networks within the cortex, and between the cortex and other brain structures. Figure 7.2 shows networks at different scales detected using fMRI in human subjects performing a series of language tasks (Caplan et al., 1998). Other networks are dynamically activated when subjects perform a series of motor tasks and the temporal evolution of these networks can be traced, as shown in Figure 7.3.

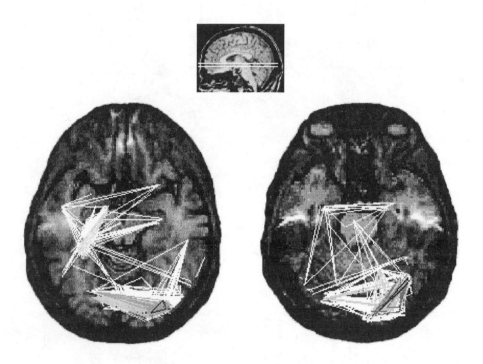

Figure 7.2. Two example slices from a whole-brain weight map computed across ten different language tasks. Inset (top row center) shows the location from which the slices were obtained. Subjects performed ten different language tasks while undergoing fMR imaging, and activation maps were computed for each task. A voxel was given a score of +1 if it was significantly activated, or 0 otherwise, and the resulting six activation maps were summed to produce a weight map indicating the number of tasks in which each voxel participated. All pairs of voxels with a given weight value were connected by lines, to indicate the level of functional connectivity across tasks (where darker lines indicate stronger functional connectivity). The strongest coupling was found at the smallest spatial scales (black), while larger regions (progressively brighter lines) appeared at lower levels of functional connectivity.

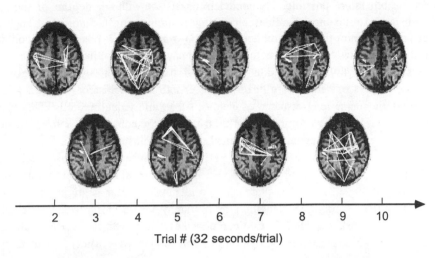

Figure 7.3. Single-slice weight-map computed at various points in time across six motor tasks. In this case, significance maps were computed for each usable block of motor activity (n = 9) during the motor tasks. Weight maps were then computed, per Figure 7.2. We display clusters for a single slice only at the highest threshold, computed over each block in the experimental paradigm. Fully connected networks were recovered at different spatial scales at all points in the experiment. Moreover, these networks changed substantially over time, suggesting dynamic organization of intermediate levels of brain structure. As new technologies for brain mapping advance, a wealth of information about the spatiotemporal organization of the neocortex is appearing. Making sense of this knowledge requires a framework to help organize the findings, especially with respect to intermediate levels of network organization and scaling across levels. Such a framework can also be useful in data interpretation, formulating new questions and guiding further investigations. The model we describe next provides just such a framework.

7.3   Network of Networks (NoN) Model

7.3.1   Architecture

The brief summary above vastly oversimplifies current understanding about the structural and functional scales of spatiotemporal organization in the cortex. From this and other evidence, however, we believe that multiple levels of spatial and temporal organization exist between single neurons and the entire cortex. Clearly the existence of such intermediate scales of organization calls for a model that can provide a framework for investigating and understanding such phenomena. The fundamental assumption of the NoN model is that the cortex – a large, immensely complex network – is organized at multiple scales, built up from self-similar functional units. Neurons cluster to form small networks, which in turn cluster to form larger networks, and so on. This notion has been suggested previously (Mountcastle, 1978), and various candidate functional units have been considered, including cortical columns, but others are possible. The model is consistent with the notions of translational invariance and specialization.

An illustration of the NoN architecture is shown in Figure 7.4 and is based on the described properties of neocortical anatomy and physiology. The model assumes that there is a high density of connections locally and sparse connectivity at a distance. In this way, clusters of neurons are linked together into larger clusters (i.e., functional networks) via sparse long-distance connections from individual neurons. This clustering continues up to the entire network. Driving of the cortex comes from outside inputs (e.g., sensory inputs, other cortical inputs, brainstem modulatory inputs). Locally, the computations are assumed to be highly recurrent, with substantial local feedback connections. Critically, the properties of individual neurons are maintained at all levels of organization (i.e., there is no population or scalar averaging involved). To limit confusion among levels within the hierarchy, we will refer to the entire system as the "network", which is in turn composed of "clusters" (i.e., smaller, sub-networks), with individual neurons providing the lowest level of (cluster) organization.

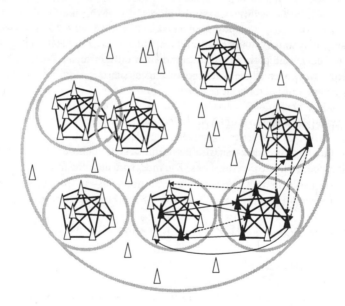

Figure 7.4. A representation of the NoN model. Sixty-four neurons are shown (triangles), forty-nine of which are grouped into seven clusters of seven neurons each. In this schematic, the neurons represent the lowest order (or trivial) clusters (open triangles = local neurons, filled triangles = projection neurons), the clusters of seven neurons represent first order clusters, and the network as a whole represents a top order cluster. The lines indicate functional connectivity at a given point in time, with heavier lines indicating stronger connections. The interconnected group of clusters at the bottom right represents a level of clustering between the individual clusters and the entire network. The boundaries of these clusters and sub-/super-clusters can overlap (upper left) and change with time. Only some neurons and connections are shown for simplicity.

The boundaries in the NoN are determined by the connection patterns, the sign and strength of synaptic weightings, and regional source(s) of inputs. Some neurons (Figure 7.4, open triangles) have only local connections, while other neurons have both local and long-distance connections (Figure 7.4, filled triangles). In principle, the boundaries between clusters are not discrete. They vary in time and may be continuous in space (down to the level of individual neurons). Plasticity in the neural connections is ongoing, and is assumed to occur at all temporal scales – ranging from short-duration, dynamic effects (e.g., neuromodulatory inputs), to long-term memory effects (e.g., stable modifications of synaptic efficacy). As such, plasticity mediates the changing functional groupings required for dynamic control, as well as for the development of higher-level representations built up by learning to connect two or more lower-level features – known as features of features. Thus, plasticity gives the model the capacity to continually redefine its structure, on time scales corresponding to the time constants of the various forms of plasticity.

Finally, the model does not presume any particular topological form short of networks binding together to form NoNs. A two-dimensional locally connected sheet of attractor networks is one spatial instantiation of a NoN model (Anderson and Sutton, 1997). However, more complex configurations are possible, including (1) higher dimensional configurations such as cubes and hypercubes, and (2) layered structures, for example, projection systems (Sutton and Anderson, 1998). The latter structure is more akin to the architectures of traditional layered neural networks and would be necessary for incorporating links to subcortical structures such as the thalamus and striatum.

7.3.2   Model Formulation

The NoN model and its predecessors and variants have been developed and analyzed in detail elsewhere (Sutton et al., 1988a; Anderson and Sutton, 1995; Sutton and Anderson, 1995). While the various implementations vary in their computational features, the core model requires the selection of mathematical formalisms for three main model components: neurons, clusters, and inter-cluster connectivity. These components define three levels of organization, and from this Sutton (1988) shows how to generalize to the n-level case. (In essence, generalization simply involves replicating the inter-cluster connectivity rules from one spatial scale to the next, preserving neuron autonomy throughout the entire network.)

The first component, the neurons of a NoN, are typically chosen to be binary state, integrate-and-fire, and all-or-none in nature. The precise selection is often made based on the availability of tools for analyzing systems of such neurons (e.g., McCulloch and Pitts, 1943), and on the level of neuronal simplification deemed acceptable.

The second component of a NoN model is the type of cluster to invoke for the first level of organization. This choice defines both the dynamic nature of the cluster, as well as the way in which neural plasticity is incorporated. Common choices are attractor neural networks such as Hopfield networks (Hopfield, 1982) or Brain-State-in-a-Box (BSB) networks (Anderson, 1993), wherein memories are encoded as stable states of the network. However, it should not be construed that the use of a specific model (e.g., BSB networks) implies that the neocortex is actually composed of these networks. It is the general architectural principles of clustering, nesting, and parallelism that are important. It is also not necessary that clusters in the NoN behave as any particular type of network (e.g., an attractor network) in order for the NoN to function.

The final component of a NoN model that must be specified is the inter-cluster connectivity. Critical to the NoN model is that inter-cluster connectivity remains vector-based – that is, over a set of connections – thereby preserving the independence of neurons at all spatial scales. Per assumption, the inter-cluster connectivity will be sparse, and generally weaker than within-cluster connectivity. When one network is in a particular state, the pattern of neural activation (i.e., the state vector) will excite patterns (states) in all connected networks to varying degrees.

7.3.3   NoN Properties

The simple process of defining a NoN model in the previous section gives rise to several interesting and unique functional properties, which we briefly outline here.

The NoN model exhibits scaling across time in addition to across space. Single neurons in the network quickly make computational decisions (e.g., to fire), while clusters take longer to make such decisions (e.g., to reach an attractor). Moving up the organizational ladder to the entire network, even longer time constants of network operation are found. Often coupled to this increase in time constant is an increase in response variability. This larger response variability at the level of the entire network is an unusual property for artificial neural networks, and is more in line with human behavioral data than traditional neural networks, wherein the averaging across large numbers of neurons results in highly reproducible responses, barring the injection of random noise (Luce, 1986).

Another unique property of the NoN model is that learning becomes easier as training progresses, rather than harder. Most neural network models learn more easily at first, and have more difficulty late in learning, eventually reaching a point of catastrophic interference whereupon all previous learning is suddenly "forgotten". The NoN model obtains the opposite learning behavior via its modularity. While individual clusters are able to learn independently, learning also occurs at the site of interference patterns (see Figure 7.5). Interference patterns are obtained only when the cluster response becomes nonlinear, which in turn happens only when there has been sufficient learning of the states in question. Thus, learning may be relatively difficult at the beginning of network learning since pattern amplitudes are small, but that learning will occur more rapidly when the network has already learned enough overall to (1) transfer information (patterns) effectively between modules and (2) respond strongly enough to drive critical modules into the nonlinear regime and thereby form a boundary. If a great deal is learned, learning again becomes difficult. This qualitative pattern of learning is also closer to the pattern seen in human learning than that of typical neural networks. Interestingly, the speed of boundary formation could serve as a match/mismatch indicator. Such a process could provide a built-in cognitive ability of sorts – much like the notion that one has more difficulty perceiving something unless one expects it.

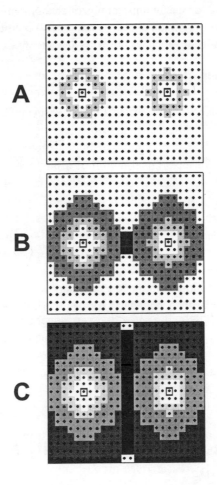

Figure 7.5. Illustration of an array of clusters in a NoN, following the external activation of two of them (Section B, left and right). Each dot represents a single cluster; grays indicate the time at which a network reached an attractor (darker colors imply later activation). Notice how the activation spreads in waves from the source over time, thereby spatializing the input. A boundary (or feature of features) forms at the point of contact due to nonlinear pattern interaction within the local clusters (black).

Finally, the NoN model allows for self-organizing behavior in time and space when processing with multiple attractors within and across networks. The dynamics of individual clusters provide the necessary component for boundary formation, and the development of such boundaries is itself self-organizing. Moreover, the NoN also provides ongoing background associations – a hierarchy of associations in place ready to receive inputs. The hierarchy of associations are the higher-level memories, or features of features generated by interference patterns as in Figure 7.5C. The ongoing processing has similarity to a dynamical function generator, continually searching a problem space, allowing for spatiotemporal regression, system control, expectation based gating, generation-recognition systems, and even reverberation, as hypothesized for memory consolidation (e.g., Wilson and McNaughton, 1994).

7.3.4   NoN Contributions

The development of the NoN is one serious attempt – several others appear in this book – to tease out some of the fundamental principles employed by the brain and the cortex in particular. The first important contribution provided by this model is a mathematical foundation for the original notion of Mountcastle (1978) that neurons organize across spatial scales. The analysis of this mathematical structure comprises a theoretical component of our NoN model (Sutton, 1988; Sutton et al., 1988b). Second, even in its most rudimentary form, the NoN possesses spatiotemporal characteristics with unusual properties. The two core properties are: (1) the notion that modules are dynamic, and (2) the ability to account for spatiotemporal scaling phenomena. Module dynamics are outlined in Section 7.3.1, and range from short-term modulatory effects to long-term memory-related changes. These dynamics are critical to the model, particularly to boundary formation (i.e., higher-level learning) within the model. The spatiotemporal scaling characteristics of the model arise directly from the architecture – simultaneously emphasizing neuronal clustering as well as the autonomy of individual neurons. This architecture provides a framework for thinking about phenomena ranging from individual neurons to the entire cortex, a span of some eleven spatial orders of magnitude.

7.4   Neurobiological Predicatability and Falsifiability

In this section, we briefly explore the properties of the cortex predicted by the NoN model and begin to explore the testability of the model itself. Testability is especially important for grand-scale theories of the brain, as it is relatively easy to develop a theory that accounts for data but cannot be refuted.

Although the NoN model is still at a relatively early stage of development, we outline here a number of validity tests of the NoN model. These are hypotheses generated by the model for which evidence is limited or does not yet exist, but if refuted would exclude various aspects of the NoN as a model of neocortical processing. We feel it is critical that such tests be made explicit for any theory of the brain, so as to provide a framework from which experimental paradigms can be developed. Some predictions for related models have been described previously (Sutton, 1997). Here, we list those that are relevant to the NoN model:

- Similar spatiotemporal features of networks exist across scales.

- Cluster boundaries change as the underlying neuronal synaptic connectivities change. Simulations of particular networks could be employed to determine whether clusters on smaller spatial scales are more or less fluid than those on larger spatial scales.

- Spatiotemporal patterns of cluster activity constitute a key information mechanism within the neocortex. Thus, neurons that are not firing during a particular task may be contributing as much information as those that are firing. Memory patterns or responses are manifested by particular dynamic patterns distributed among different clusters. This property is ubiquitous and may challenge well-accepted maps of cortical localization.

- Disruption of long-range (i.e,. between cluster) connections will tend to affect higher-level memory functions (representations consisting of features of features) more so than the disruption of short-range (i.e., within-cluster) connections.

- Convergent and divergent pathways of vector connections between clusters will result in maximum correlations from disparate regions with an average lag of zero.

- Clusters at the point of boundary formation are more susceptible to plasticity and also will have slower cluster dynamics.

The essence of the NoN's novel predictions, however, resides in the notion of dynamic repartitioning of the clusters. Clusters are recruited, they recruit individual neurons to help out as needed, they dynamically self-assemble, and they learn to anticipate upcoming inputs. Where the model becomes useful is in considering system perturbations – a sudden change in context, a novel stimulus, an injury, or a major change in modulatory input. At that point, the model can provide a prediction of what one would expect, initiating experimental tests and subsequent theoretical modifications, as required. Critically, the predictions provided will inherently span multiple spatial and temporal scales, making it applicable to, for example, experiments involving simultaneous multiunit cellular recording and fMR imaging.

7.5   Implications for Neuroengineering

The architecture of the NoN itself has several unusual features pertinent to hardware and software design. First, computation is performed by self-similar clusters whose processing power is an exponential function of the number of clusters. Second, the total connectivity and processing speed both scale linearly with the number of clusters. Finally, the architecture can automatically reconfigure itself through use of dynamic boundary formation (i.e., learning via connections that cross cluster boundaries or dynamic enhancement/suppression of neural activity).

All of these features are significant attractions for very large scale integration (VLSI) hardware implementations of the NoN. Hardware implementation of the NoN model could be very efficient as it would simply require implementing many repeating units (i.e., clusters; Mason and Robertson, 1995). Because the connectivity scales linearly, the addition of a module requires relatively few new connections between elements. Processing speed also scales linearly with network size, so exceptionally large networks will not suffer in terms of performance. Importantly, storage capacity increases exponentially with the number of levels. This combination of features is particularly unusual for a neural network model, where connectivity typically scales quadratically with size while capacity only scales linearly with size. As a demonstration of these claims, some parallel computers have had a resemblance in architecture to the NoN (Hillis, 1985).

While there are presently only a few software applications of the NoN architecture, there are a large number of problem domains within which the model appears applicable. One application is in distributed, adaptable computing environments, such as monitoring complex communication systems, involving multiple autonomous agents (Sutton and Jamieson, 2001, 2002). Individual clusters are assigned to agents, and the relatively sparse connectivity between clusters minimizes the need for high-bandwidth communication between clusters. Another potential area of application is data mining and parallel search, where each cluster performs a search locally, and communicates results to other clusters if and when a target is found. Highly parallel computations, such as those required for protein folding applications and high throughput data analysis, would also benefit from a NoN-type architecture, particularly given its ability to consider multiple solutions simultaneously (not unlike quantum mechanical computation). Finally, fraud detection (which is an example of data mining for predictive purposes), security monitoring, and data compression are other areas of application that also appear particularly amenable to NoN attributes.

It should not be overlooked that technological advances inspired by the NoN may be used to improve the design of computational devices employed in brain studies. New computational techniques may be required to analyze complex spatiotemporal data simultaneously recorded from many different brain regions and from different levels of neural organization. An overwhelming challenge facing the brain sciences today is handling the enormous amounts of data being generated in many laboratories. Solutions to this growing problem need to foster methods in data sharing, standardization and modification. Structures with multi-level and multi-tasking capabilities, such as the NoN, may play a role.

7.6   Concluding Remarks

The NoN model is different from many abstract neural network models – being both motivated and constrained by physiological knowledge. There are many things the model cannot account for, as a result of the many simplifications made in the design. Yet, the architecture is simple, self-replicating and the dynamics allow for adaptable parallel computing with similarities to biological cortical networks. Critically, even in its most rudimentary form, the NoN provides a framework for considering the relationships across spatial and temporal scales. And, the model itself possesses spatiotemporal characteristics with unusual properties. These properties should be investigated because they may yield further insight into the mystery of how the cortex functions.

7.7 Acknowledgements

We would like to acknowledge James Anderson for many fruitful discussions, and Jeremy Caplan for his help with the fMRI and weight mapping work. Support from ONR N00014-97-1-0093, ONR N00014-99-1-0884 and NIH K21-MH01080 is gratefully acknowledged.

References

Anderson, J.A. (1993) The BSB model: a simple nonlinear autoassociative network. In: M.H. Hassoun (Ed.) Associative Neural Memories: Theory and Implementation. Oxford: Oxford University Press.

Anderson, J.A., Sutton J.P. (1995) The network of networks model. Proceedings of the World Congress of Neural Networks, Vol I., pp 145–152. Washington, DC: Erlbaum.

Anderson, J.A., Sutton J.P. (1997) High performance computing and neural and physical processes: If we compute faster, do we understand better? Behavior Research Methods, Instruments, and Computers 29: 67–77.

Brodal, A. (1981) Neurological Anatomy. New York: Oxford University Press.

Brodmann, K. (1909) Vergleichende Lokalisationslehre der Grosshirnirinde. Leipzig: Barth.

Brown, T.H., Kairiss, E.W., Keenan, C.L. (1990) Hebbian synapses: biophysical mechanisms and algorithms. Annu Rev Neurosci 13: 475–511.

Caplan, J., Benson, R., Hodgson, J., Bekken, K., Rosen, B., Sutton, J. (1998) Weightspace mapping of fMRI language tasks. In: J. Bower (Ed.) Computational Neuroscience: Trends in Research. New York: Plenum Press, pp. 585–590.

Chapin, J.K., Nicolelis, M.A. (1999) Principal component analysis of neuronal ensemble activity reveals multidimensional somatosensory representations. J Neurosci Methods 94: 121–140.

Cohen, J.D., Perlstein, W.M., Braver, T.S., Nystrom, L.E., Noll, D.C., Jonides, J., Smith, E.E. (1997) Temporal dynamics of brain activation during a working memory task. Nature 386: 604–608.

Connors, B.W., Gutnick, M.J. (1990) Intrinsic firing patterns of diverse neocortical neurons. Trends Neurosci 13: 99–104.

Courtney, S.M., Ungerleider, L.G., Keil, K., Haxby, J.V. (1997) Transient and sustained activity in a distributed neural system for human working memory. Nature 386: 608–611.

Friedrich, R.W., Laurent, G. (2001) Dynamic optimization of odor representations by slow temporal patterning of mitral cell activity. Science 291: 889–894.

Georgopoulos, A.P., Taira, M., Lukashin, A. (1993) Cognitive neurophysiology of the motor cortex. Science 260: 47–52.

Gupta, A., Wang, Y., Markram, H. (2000) Organizing principles for a diversity of GABAergic interneurons and synapses in the neocortex. Science 287: 273–278.

Hillis, W.D. (1985) The Connection Machine. Cambridge, MA: MIT Press.

Hopfield, J.J. (1982) Neural networks and physical systems with emergent collective computational abilities. Proceedings of the National Academy of Sciences USA 79: 2554–2558.

Kandel, E.R., Schwartz J., Jessell, T.M. (Eds.) (1991) Principles of Neural Science. New York: Elsevier.

Luce, D. (1986) Response Times. New York: Oxford University Press.

Makeig, S., Westerfield, M., Jung, T.P., Enghoff, S., Townsend, J., Courchesne, E., Sejnowski, T.J. (2002) Dynamic brain sources of visual evoked responses. Science 295: 690–694.

Mason, R.D., Robertson, W. (1995) Mapping hierarchical neural networks to VLSI hardware. Neural Networks 8: 905–913.

McCulloch, W.S., Pitts, W. (1943) A logical calculus of the ideas immanent in nervous activity. Bull Math Biophysics 9: 115–133.

Mountcastle, V.B. (1978) An organizing principle for cerebral function: The unit module and the distributed system. In: G. Edelman, V.B Mountcastle (Eds.) The Mindful Brain. Cambridge, MA: MIT Press, pp. 7–50.

Nicolelis, M.A., Ghazanfar, A.A., Stambaugh, C.R., Oliveira, L.M., Laubach, M., Chapin, J.K., Nelson, R.J., Kaas, J.H. (1998) Simultaneous encoding of tactile information by three primate cortical areas. Nat Neurosci 1: 621–630.

Pouget, A., Dayan, P., Zemel, R. (2000) Information processing with population-codes. Nat Rev Neurosci 1: 125–132.

Rockel, A.J., Hiorns, R.W., Powell, T.P.S. (1980) The basic uniformity of the neocortex. Brain 103: 221–244.

Sanes, J.N., Donoghue, J.P. (2000) Plasticity and primary motor cortex. Annu Rev Neurosci 23: 393–415.

Shepherd, G.M. (Ed.) (1990) The Synaptic Organization of the Brain. New York: Oxford University Press.

Sutton, J.P. (1988) Hierarchical organization and disordered neural systems. Toronto: University of Toronto.

Sutton, J.P. (1997) Network hierarchies in neural organization, development and pathology. In: C.J. Lumsden, W. Brandts, L.E.H Trainor (Eds.) Physical Theory in Biology. New Jersey: World Scientific, pp. 319–363.

Sutton, J.P., Anderson, J.A. (1995) Computational and neurobiological features of a network of networks. In: J.M. Bower (Ed.) Neurobiology of Computation. Boston: Kluwer Academic, pp. 317–322.

Sutton, J.P., Anderson, J.A. (1998) System and method for high speed computing and feature recognition capturing aspects of neocortical computation. General Hospital Corporation and Brown University, US: Research Corporation.

Sutton, J.P., Beis, J.S., Trainor, L.E.H. (1988a) Hierarchical model of memory and memory loss. Journal of Physics A: Mathematical and General 21: 4443–4454.

Sutton, J.P., Beis, J.S., Trainor, L.E.H. (1988b) A hierarchical model of neocortical synaptic organization. Mathl. Comput. Modelling 11: 346–350.

Sutton, J.P., Jamieson, I. (2001) Reconfigurable network of neural networks for autonomous sensing and analysis. Fifth International Conference on Cognitive and Neural Systems, 2001: 64.

Sutton, J.P., Jamieson, I.M.D. (2002) Reconfigurable networking for coordinated multi-agent sensing and communications. In: H.J. Caulfield et al. (Eds.) Sixth Joint Conference on Informational Sciences. Research Triangle, North Carolina: Association for Intelligent Machinery: 36.

Szentagothai, J. (1977) The neuron network of the cerebral cortex. Proc. R. Soc. Lond. B. 201: 219–248.

Tanaka, K. (1993) Neuronal mechanism of object recognition. Science 262: 685–689.

Ts'o, D.Y., Frostig, R.D., Lieke, E.E., Grinvald, A. (1990) Functional organization of primate visual cortex revealed by high resolution optical imaging. Science 249: 417–420.

Van Essen, D.C., Anderson, C.H., Felleman, D.J. (1992) Information processing in the primate visual system: an integrated systems perspective. Science 255: 419–423.

Wilson, M.A., McNaughton, B.L. (1994) Reactivation of hippocampal ensemble memories during sleep. Science 265: 676–679.

Chapter 8

Towards Global Principles of Brain Processing

John G. Taylor

8.1 Abstract

A set of principles are developed to explain how general information processing is carried out in the brain. These involve sub-cortical sites to help create specific control structures to achieve active responses to inputs and to develop memory systems for more effective responses to the environment. Consciousness and thinking are regarded as top-level processes created by suitable attentionally driven brain structures which are identified mainly in posterior and anterior sites respectively. In particular a specific neural model, the CODAM model, is developed to explain consciousness. Experimental evidence for the principles and their neural adumbrations are briefly surveyed.

8.2 Introduction

The brain presents an enormous challenge to our understanding. This has developed considerably over the previous decade with the aid of brain imaging machines, with remarkable images of internal brain activations during subjects' completion of various cognitive tasks (Posner and Raichle, 1994). Those involved in such an enterprise are now to be regarded as "internal phrenologists". Various attempts have been made to bring together the understanding being gained from analysis of specialized regions or of brain response for specialized tasks, and several reviews have recently appeared combining these features (Mesulam, 1998; Cabeza and Nyburg, 1997; Damasio,

1994; Halgren, 1994). There are also more specific proposals coming from the neural network community (Hecht-Nielsen, 1998; Freeman, 1998). Consciousness has become of interest to many neuroscientists and neurophilosophers (Baars, 1988; Dennett, 1991; Crick, 1994, Taylor, 1999).

In order to extract principles from this "embarrassment de riches" I will begin by considering the nature of such principles which we should expect to find to enable coherence to be achieved. I conclude that we need to work at different levels and bring information across these levels to enable understanding to arise. I then consider how new understanding is to be seen arising at the highest, network level through structural modelling of brain imaging results. How this is helping us to piece together the disparate networks is considered in the following section. A particular experimental paradigm is then considered, the waterfall effect, which is used to indicate the network of sites involved in high level processing. The "highest" level of all, that of consciousness, is then considered in more detail in the following section, in terms of a three-stage model and more specifically the CODAM model based on a control approach to attention (Taylor, 2001, 2002a). Global brain principles are then developed for whole brain processing in a later section, especially based on attention control processes. After a discussion on neural systems underpinning "thinking", the chapter concludes with a discussion.

8.3   What Could Brain Principles Look Like?

If one considers "hard" science, there are beautiful examples of basic principles on which a particular branch of such science is founded. Thus, special relativity can be based solely on the principle: "The velocity of light is always the same, whatever movement of the apparatus occurs when it is measured". This is such a simple principle, yet from it can be predicted the subtlety of mass increase and space contraction with increasing speed, the twin paradox and many other unexpected features. The Lorentz group $O(3, 1)$ and its representations as corresponding to the observed elementary particles thereby take meaning in the physical world, and not just in the brains of pure mathematicians.

Another example is that of quantum mechanics, which can be similarly founded on two basic principles:

1.   Dynamics: described by the Schrodinger equation for the wave function, $id\psi / dt = H\psi$, where H is the energy of the system with wave function ψ.

2.   Measurement: given by the probability of a measurement a of any variable being the quadratic expression $|<\psi|a>|^2$, where $|a>$ is the eigenstate of the system with value a of the measured quantity, and $<\psi|a>$ is the overlap of the system state ψ and the eigen state of measurement, $|a>$.

Again an enormously rich and complex world arises from these simple postulates. In both cases of special relativity and quantum mechanics the postulates have yet to be found wanting experimentally.

What is the comparable situation for the brain? It is composed of neurons, so is based on the fundamental neuronal equations (with shunting) of Hodgkin and Huxley for the membrane potential:

$$\tau_i \, du_i / dt = - (u_i - E_o) + \Sigma_j \, (u_i - E_\alpha) \, g_{ij} \, f_\alpha \, (u_j) \, , \qquad (1)$$

where the label α denotes the particular ion channel, τ_i the time constant of the ith neuron and the quantities g_{ij} are the conductances of the synapses coupling neuron j to the ith neuron. Equation 1 is an enormous simplification leaving out many of the features which may be crucial to obtain suitable sensitivity in response:

- spiking (which can be included by suitable choice of the output function $f_\alpha \, (u_j)$ in Equation 1;

- cell geometry (which can be incorporated by extending the membrane potentials so as to be multi-component vectors, each component arising from a compartment of the cell, with suitable linking terms for current flow along the dendrites);

- stochastic transmission (which needs extension of Equation 1 to allow the conductances g_{ij} to be stochastic, chosen independently at each time step from a given distribution);

- adaptive synapses, in which resources are depleted by activity, and various channel variables take account of resource recuperation (as can be given in terms of the facilitatory or depressive synapses considered recently, Markram, 1998);

- effects of neuromodulators, such as dopamine, serotonin, noradrenaline (which can be modeled by additional multi-channel and second messenger additions to Equation 1).

These various features combine to make even more complex the patterns of response given by neurons obeying Equation 1. These complexifications can be regarded as similar to those arising when more complex structure is imposed on quantum mechanics, such as arising from the discovery of particle creation and annihilation at particle accelerators, or the need to introduce superstrings in order to unify quantum mechanics and gravity. Such increased complexity is essential to understand the physical world as it is observed now. However, there was a historical progression from "simple" quantum mechanics for fixed numbers of particles to the more complicated quantum field theories describing particles at high energies. I will take the same "softly softly" approach, and start with the simple forms of neurons as exemplified by Equation 1, and only subsequently shall I consider possible modifications which the above bullet points might bring to bear; I will have no space to do that here.

The neurons activate each other through long axon tracts joining otherwise distant parts of the brain (together with local neuronal interactions for the outputs of inhibitory interneurons). The separate regions are observed to activate each other when seen through the eyes of brain imaging machines, PET, fMRI, MEG or EEG. The temporal and spatial patterns of activity are now becoming exposed to our increasingly penetrating gaze. How can we attack the problem of what are the global principles at work as a subject solves a cognitive task?

One way would be to follow the tripartite division of analysis suggested by Marr (1982), into:

1. a computational theory (what is the goal of the computation, and the strategy being used),

2. a representational/algorithmic theory (what representation of the information and resulting algorithm is being used to solve the task),

3. a hardware implementation (what hardware realizes physically the algorithms carrying out the strategy).

This division has proved attractive in attacking the problem of vision, but meets difficulties when facing the more general task of uncovering global brain principles. In particular the results arising from brain imaging very heavily involve the detailed neuronal implementation as well as complications arising from blood flow (Marr's third part), as well as the first two parts, in an intermixed manner. However, we will use this division behind our discussion to allow us to abstract from the specific brain patterns the more general principles being implemented.

What is more problematic is how we relate the basic underlying neuronal equations, say of form (1), to the overlying functionality of brain modules. This is precisely the problem presented in going between levels 1 and 3 of the approach Marr referred to above. The problem for the brain is that the neuronal implementation of level 3 is on an "atomic" scale, whereas the computational goals and strategies used, as well as important components of the algorithmic implementation are at a more, if not total, global scale. Thus we have to work down from the most global level of functionality if we wish to understand all possible components that are involved. We will do that by turning to more global aspects of brain processing as observed by imaging machines before relating this activity to that at the atomic, neuronal, level. Only then do we turn back to the highest levels of attentional control and consciousness to attempt to extract brain-processing principles.

8.4   Structural Modeling

Initially PET and fMRI studies uncovered a set of active brain sites involved in a given task. This is usually termed a "network", although there is no evidence from the given data that a network is involved but only an isolated set of regions. It is possible to evaluate the correlation coefficients between these areas, either across subjects (as in PET) or for a given subject (in fMRI). There is great interest in using these correlation coefficients to determine the strength of interactions between the different active areas, and so uncover the network involved. Such a method involves what is called "structural modelling", in which a linear relation between active areas is assumed and the path strengths (the linear coefficients in the relation) are determined from the correlation matrix.

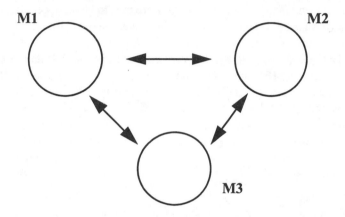

Figure 8.1. A set of three interacting modules in the brain, as a representative of a simple structural model. The task of structural modelling is to determine the path strengths with which each module affects the others in terms of the cross-correlation matrix between the activities of the modules.

Activity levels z_i ($i = 1, 2, 3$) of the modules of Figure 8.1 are assumed to satisfy a set of linear equations:

$$z_i = \Sigma a_{ij} z_j + \eta_i \tag{2}$$

where the variables η_i are assumed to be independent random noise variables. It is then possible to calculate the coefficients a_{ij} from the correlation coefficients $c(i,j)$ between the variables z. This can be extended to comparing different models by a chi-squared test so as to allow for model significance testing to be achieved.

An important question is how to bridge the gap between the brain imaging data and the underlying neural network activity. In particular, the interpretation of the structural model parameters will then become clearer. One is by the mean field approximation, so that the particular label of a neuron in a given module is averaged over. The averaged equations for the membrane potentials $u(i,t)$ for the mean neuron in the ith module at time t then results as:

$$\tau du(i, t) / du = - u(i, t) + \sum_j c(i, j) f [u(j, t - t(i, j))] \tag{3}$$

where $c(i, j)$ is the mean connection strength between the modules i and j and $t(i, j)$ the associated time delay. In the limit of $\tau = 0$ and for linearly responding neurons with no time delay there results from Equation 1 the equation:

$$u(i,t) = \sum_j c(i,j)\, u(j, t) \tag{4}$$

which is the structural equation of the form of Equation 2 now being used to analyze the data of PET and fMRI. The path strengths are thus to be interpreted as the average connection weights between the modules. Moreover the path strengths can be carried across instruments, so as to be able to build back up to the original neural networks. This involves also inserting the relevant time delays as well as the connection strengths to relate to MEG data. It should be added here that PET and fMRI measure blood flow/oxygen level and are not directly measuring neural activity. Thus the structural model for data from these instruments should be derived from Equation 4 by relating the measured features of blood/oxygen to the neural action (Taylor et al., 2000b). We see here a very clear entry into the arena of the specific implementation of the processing in wetware. Marr's third level cannot be neglected in the analysis of brain imaging data.

8.5   Static Activation Study Results

There are many psychological paradigms used to investigate cognition. A number of these have been used in conjunction with PET or fMRI machines. These paradigms overlap in subtle ways so that it is difficult at times to "see the wood for the trees". To bring in some order we will simplify by considering a set of categories of cognitive tasks. We do that initially along lines suggested by Cabeza and Nyberg (1997), who decomposed cognitive tasks into the categories of:

- attention (selective/sustained)

- perception (of object/face/space/top-down)

- language (word listening/word reading/word production)

- working memory (phonological loop/visuospatial sketchpad)

- memory (semantic memory encoding and retrieval/episodic memory encoding and retrieval)

- priming

- procedural memory (conditioning/skill learning)

So far the data indicate that there are sets of modules involved in the various cognitive tasks. Can we uncover from them any underlying functionality of each of the areas concerned? The answer is a partial "yes". The initial and final low-level stages of the processing appear more transparent to analysis than those involved with later and higher level processing. Also, more study has been devoted to primary processing areas. Yet even at the lowest entry level the problem of detailed functionality of different regions is still complex, with about 30 areas involved in vision alone, and another 7 or 8 concerned with audition. Similar complexity is being discerned in motor response, with the primary motor area being found to divide into at least two separate subcomponents. The details of these many areas are slowly being unraveled.

8.6 The Motion After-Effect (MAE)

Let me give an example of what fMRI may look like and the sort of networks that can be involved. The MAE effect is sometimes called the "waterfall effect", since it occurs if you look at a waterfall for some 20 seconds and then look at the rock-face on either side; this appears to be moving upwards (and does so for about 9 seconds). We used fMRI to study the effect since it allowed determination of the timing of the emergence and continuation of activity time locked to the motion-after effect period (Schmitz et al., 1998; Taylor et al., 1998). The possibility of observing such time-courses depends crucially on using a suitable paradigm to enable the timing to be specified. We used a moving stimulus (in our case a set of moving bars) which ceased their motion at a certain time so that we had an offset point from which to time any changes of activity. It was also necessary to compare with activations brought about by exposure of the subject to a control condition, such as the bars moving up and down. For that stimulus there would have been no after-effect. This was the basis of the method used in Tootell et al. (1995) (although with a different stimulus composed of moving and static rings), and was used more extensively in our study. We searched for activations, especially in prefrontal sites, which were timed exactly with the offset of either bars moving in one direction (we only used bars moving down in this study), or up and down successively. Any regions which had activation which commenced as the stimulus movement ceased at the end of the motion down period, and which then decayed over about the expected time for the MAE experience, was a candidate for the network of areas subserving the MAE experience itself. The signal after the one-way movement, as compared to that in the up-and-down case, gave the net MAE effect. That characteristic was the main signal for which we searched in the fMRI data.

There were also other characteristics of signals expected from regions involved in the overall processing. Thus we expected to see sites involved in analysis of the static bars during the period just after the downward motion. This period itself can be split into one during which the MAE is being experienced and a period in which there are only static bars, together with the fixation cross, being observed and experienced. In order to achieve the highest attention during the whole period of exposure of the subject we did not tell them not to expect to experience any MAE after the up-and-down motion period. So there was also the putative MAE period after the up-and-down motion period. We were interested in those areas which were especially active during one or more of downward MAE, putative up-and-down MAE and at other times when the bars were static as well as during either motion period.

Our main finding (Taylor et al., 1998) is that there are time courses in both posterior and prefrontal cortical areas which differ significantly across these periods and across the different areas. Some sites (such as the frontal ones in BA 44, 47 or the cingulate gyrus and more posterior in BA 40L) were only active during either of the possible MAE periods (for a few seconds after the cessation of either the movement down of the bars or that in both directions). Other time courses (in BA 40R, for example) were only or mainly active during downward-caused MAE. Yet others (such as in MT/V5) were active both during either motion period and during the downward MAE. These results indicate that there are different networks involved in different stages of the total experience. Some areas have different components involved in different stages and very likely in different networks.

The overall picture uncovered of the modus operandi of the brain areas involved in processing the MAE is one of different networks of sites involved in different tasks. These are:

a) responding to the static or to the moving stimuli,
b) checking for the occurrence of MAE,
c) creating the experience of MAE itself.

The networks involved in these various tasks are found to be overlapping to a certain extent. Thus checking MAE and creating the experience both involve BA 44, BA 40, cingulate gyrus and BA 46/47/10. This was supported by calculation of the correlation matrix of activities between the different sites; the correlation values were large (>0.5) for the separate anterior and posterior nets with only one or two key areas joining the two networks together, and then only with correlation coefficients at about 0.5. The overall picture of this set of networks is shown in Figure 8.2. This indicates that there are two main networks which communicate to each other via MT and BA 40.

Which of these is involved in the creation of awareness of the MAE experience itself (task (c) in our above list)? Up until now MT was regarded as the site of such experience. Our data indicate that another candidate for the creation of this experience is also BA 40. Use of independent component analysis (Karhunen et al., 1997) singles out two separate signals, one especially involved with MT as an overall motion detector, and the other contained mainly in the anterior sites BA 44/40/47 (not shown explicitly in Figure 8.2 but part of the anterior network) and CG (Taylor, Schmitz and Fellenz, 1998). The second independent component is exactly timed to the MAE signal. Thus it would appear as if the experience of the MAE is based more frontally than posteriorly, although there is a part of the MAE signal detectable in MT (as the first experimental results of Tootell et al., 1995, on this showed convincingly). Thus consciousness is created in a network of sites, although not all of the cortex is directly used in such activation. Early cortical processing, such as in V1, does not

have a distinctive MAE signal, so is not as involved (it is not in other early processing, such as orientation sensitivity (He et al., 1996).

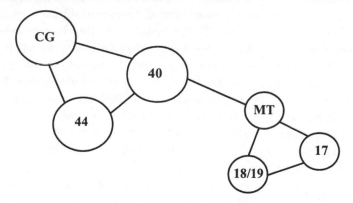

Figure 8.2. The two networks: BA 40, 44 and cingulate gyrus (CG) and BA 18/19, MT and 17,
with the path strengths greater than 0.5 shown by connecting lines. MT and BA 40 are the two
key areas which carry information from one network to the other.

8.7 The Three-Stage Model of Consciousness

So far we have seen that there are areas of the cortex which are involved in creating
the experience of awareness of MAE, as well as those which are not directly so
involved but are functioning in a preprocessing mode. This is the case for V1, for
example. This dissociation of consciousness is supported by a number of experiments, indicating that areas of cortex are solely involved in preprocessing (He et al.,
1996; Taylor et al., 2000a, to be discussed below). There is also a network of areas
with a strong correlation with the MAE experience, for example. Finally there are
"checking" or more general attentional control areas (mainly in the frontal cortex)
which are involved at a higher level of control.

   We arrive thereby at the three-stage model of consciousness creation (extending
that in Taylor et al., 2000a) of Figure 8.3, in which the third stage is involved with
higher-order control processes for attention and sequence processing (which may be
independent of consciousness):

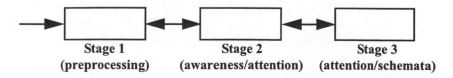

	Stage 1	Stage 2	Stage 3
	(preprocessing)	(awareness/attention)	(attention/schemata)

Figure 8.3. The three stages of the emergence of awareness. Stage 1 is only involved with preprocessing, as in early visual or auditory cortex. In stage 2 simple awareness of the environment emerges through low-level control structures for attention, whilst in stage 3 higher-level
control of attentional processes occur to create schemata for sequences of action processes.

The relation between the three stages of Figure 8.3 and the two networks of Figure 8.2 is still to be worked out. Further evidence was presented for the three-stage model of Figure 8.3 in Taylor et al. (2000a) as applied to awareness and attention of auditory stimuli. Subjects were exposed to sequences of syllables and had to listen to them under the three conditions of (a) inattention, when they were required to detect flickering of visual inputs, (b) awareness, when they listened to the syllables with no other distractor but were not required to process them strongly, and (c) attention, when they had to count the number of times the syllable /ta/ was spoken. In proceeding from condition (a) to (c), with increasing attentional load, increasing numbers of cortical areas became active as detected by fMRI. These areas were in expected regions, with those for inattention involving primary auditory areas, those for awareness sited more posteriorly in the inferotemporal cortex (mainly left) and finally the higher-order attentional control sites were predominantly in the frontal cortex. Nor were there strong changes of activity in the areas still being used in the higher attentional load cases as compared to the lower attentional ones. In all the sets of areas observed under the conditions (a), (b) and (c) can be identified with the modules of stages 1, 2 and 3 respectively of the three-stage model of Figure 8.3.

The three-stage model has a further extension, supported by considerable psychological and brain imaging data that helps to fill in some of the details as to how awareness arises in the first instance in stage 2 modules. This involves the inclusion in the stage 2 modules of cortical areas specialized by the long duration of their neural activity. These are the "slave" working memory sites proposed as part of the model of working memory of Baddeley and Hitch (Baddeley, 1986). These sites have now been observed directly by brain imaging techniques (Smith et al., 1998). It has also been suggested (Taylor, 1998) that the cortical sheet can be approximated as a continuous sheet and then neural field models can be considered which predict the existence of "bubbles" of continued activity. The more precise characteristics of the neural systems needed to create phenomenological awareness were explored in Taylor (1998, 1999, 2000).

The argument of Taylor (1998, 1999, 2000) can be summarized as follows:

1.   Consciousness arises at the lowest level by recurrence of neural activity, thereby giving the activity extended duration at the same time as allowing to settle into suitable most effective states to remove ambiguity in interpretation (if there were any). The extended temporal duration is important to allow time for the activity to win a form of competition against other similar longer-lived activity fighting to gain the arena of consciousness.

2.   Part of the effects of gaining awareness is that a percept now has access to the rich panoply of the higher-level cortex, especially frontal sites with access to global report, rehearsal and transformation. This process is one expected to be taking place at the buffer sites of working memory in posterior multi-modal cortex.

3.   These working memory buffer sites are well connected to the hip-
     pocampus, an essential organ for laying down of memory. With-
     out a hippocampus no long-term episodic memory can be
     attained; with it is developed the sense of self through connections
     between suitable frontal sites (especially in BA 10 in the mesial
     frontal region of cortex) and those of storage of episodic memory
     very likely in mesial posterior sites such as the retrosplenial gyrus
     and the precuneus).

4.   The exact sites of appearance of conscious awareness is not pres-
     ently known. In any case that will very likely vary as different lev-
     els of attention modulate the levels of activity in earlier cortical
     areas, as now observed by many brain imaging and single cell
     experiments although, importantly, effects on time durations are
     not known. However there may well be a form of "on-line" con-
     sciousness which involves a fleeting level of awareness that can
     be caught if attended to but in many situations is not. That this
     low-level awareness is not remembered can be explained by it not
     arising in the buffer working memory sites and so not having
     good access to the hippocampus.

These arguments lead to adding a fourth level of consciousness, thereby extending the model of Figure 8.3 to that of Figure 8.4. The original stage 2 has been split
into an on-line version (stage 2a) of which there is little memory, and a longer term
one (stage 2b). Phenomena such as the attentional blink allow the dissociation of
stage 2 level of awareness to be experimentally probed.

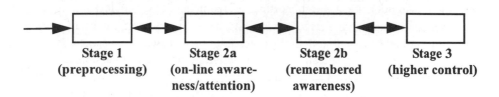

Stage 1	**Stage 2a**	**Stage 2b**	**Stage 3**
(preprocessing)	**(on-line aware-**	**(remembered**	**(higher control)**
	ness/attention)	**awareness)**	

Figure 8.4. The four stages of the emergence of awareness. Stage 1 is only involved with preprocessing, as in early visual or auditory cortex, as in Figure 8.3. Stage 2 of the earlier model
has been split into "on-line" simple awareness of the environment, which is poorly remembered, and stage 2b, which arises over a longer period of time and is remembered in episodic
memory. In stage 3 higher order control processes allow memory of the conscious experience
to be built as well as filter and control input and later processing, as in the original model of
Figure 8.3.

The relation between the two stage 2 levels of consciousness are of great interest;
the present position is unclear, with controversy arising over, for example, if color
arises in V4 or there is awareness of activity in V2 (Pollen, 1999; Crick and Koch,
1998a, 1998b). The model of Figure 8.4 may be too sharply divided to take account

of the true state of affairs, and a more gradual approach to true awareness is occurring in proceeding from modules which are definitely stage 1 to those which are definitely stage 3. The model of Figure 8.4 can be extended to account for that by having a sequence of levels between the two extremes, with a sequence of neural "bubbles" of increasingly long time durations, as already described (Taylor, 1998).

Let us now turn to the higher level of stage 3 processing. There is increasing understanding of the source of attentional activation in the modules of this stage. There has been observed even more extended duration of activity in frontal areas than that in the posterior slave sites. The cause of this activity has been suggested as arising from the functioning of "loops" of activity flowing through the cortex-basal ganglia-thalamus-cortex loops (Taylor, 1995), for which there is strong neuroanatomical support. The resulting so-called ACTION network can be seen to provide a platform for explanation of the numerous functions performed by the frontal lobes, such as the storing of goals, inhibition of prepotent responses, planning, thinking, and control of social responses. We thus extend the global circuitry of Figure 8.2 to consist of the following:

- stage 1 modules: preprocessing sites (up to semantic level) in the posterior cortex;

- stage 2 modules: buffer "working memory" sites, such as the phonological store, the visuospatial sketchpad, the body matrix, and so on;

- stage 3: a set of coupled ACTION nets in the frontal cortex, including their associated basal ganglia and thalamic sites.

We should also not neglect the subcortical (thalamic and other nuclei) support systems for both the stage 1 and stage 2 modules. They also play their important roles to support the complete functioning of the preprocessing and awareness stages.

The possibilities of dissociation between activities in these various sites is clearly considerable, caused by too heavy a load, damage to various sites, or drug and other biochemical interferences. This is an indication of the complexity of consciousness itself, so that it is thereby not easy to give a definition of consciousness. It is composed of a number of components of which one division can be into the parts:

- posterior or passive;

- driven by additional activity from anterior cortex;

- self, driven by further activity in hippocampus and orbito-frontal cortex as social rules and memories of self;

- emotional, contained also in mesial cortical and limbic sites such as amygdala.

Each of these major subdivisions can themselves be subdivided into further components, as seen from analysis of the effects of brain defects on experience as well as from more detailed brain imaging studies.

That there is a unity of the totality of consciousness, as we experience it in our normal waking lives, is a miracle, given the above divisibility. However, there must be global control circuits able to give unity to what would otherwise be a confusing chaos of internal activity. This will shortly be developed as part of the CODAM model as creating the important component of consciousness associated with the so-called pre-reflective self. This is the component involving the sense of "what it is like to be" (Nagel, 1974). It is such inner experience that gives ownership to the multifold contents of experience described so far. Without that inner sense it is not satisfactory to claim there is "anyone at home". How can such an inner sense arise? I turn to give a tentative neural answer to that in the next section.

8.8   The CODAM Model of Consciousness

The problem faced in creating a neural model of inner experience is to give a detailed specification of what is needed to be achieved: what is this inner sense like? It is well accepted now that attention is moved by a higher-level neural system separate from the regions in earlier cortex to which attention is being directed. Thus attention is properly described as a control system, and has been approached as such by engineering control methods (Taylor, 2000). An engineering control approach allows a unified view of attention and new understanding to be gained as to its use (Taylor and Rogers, 2002). In the control model, an observer (an estimate of the attention state of the system) contains a buffered efference copy of the controller signal. This is used to achieve more rapid updating of the movement control signal than if external feedback had to be waited for. Such a copy will not be bound in any attention-based manner to the content of consciousness, since that can only be present on feedback from lower cortical sites. Such an efference copy signal will therefore not have any content. It can, however, be identified with the experience of "ownership", that of the about-to-appear amplified input that is being attended to. The corollary discharge signal will grant immunity to error through misidentification of the first person pronoun (Shoemaker, 1968). This will occur if the corollary discharge acts as a "sentry at the gate" of the working memory buffer for the input. The discharge then only lets onto the buffer what it has been told to by the inverse attention controller. As such it inhibits all other possible entrants to contentful consciousness. This occurs for the brief period before the attentionally amplified input from the sensory cortex arrives. The corollary discharge is then inhibited in its turn. Such complex processing is supported by the siting of much of the attention control structures nearby in the parietal lobe, singled out recently as crucial for consciousness to arise (Taylor, 2001). A possible control model for this is shown in Figure 8.5.

Following many experimental results, we site the IMC in the superior parietal lobe, the working memory buffers in the inferior parietal lobe, the monitor in the cingulate, the goals module in the prefrontal cortex or superior colliculus and the attentionally modulated cortex in the sensory cortices.

There results the CODAM model (Taylor, 2000, 2002a, 2002b):

> *The pre-reflective self is identified as experienced as the corollory discharge of the attention movement control signal residing briefly in its buffer until the arrival of the associated attended input activation arriving at its buffer inhibits the former signal.*

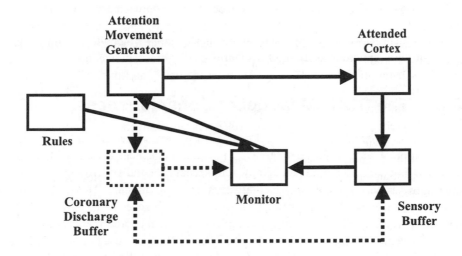

Figure 8.5. A simple attention movement control model (shown in bold lines), composed of the attended cortex (containing activity representing an attended input), an attention movement generator, a rules module (for either top-down or bottom-up control of attention) and a monitor (based on the error between the required attention state and that occurring as determined by a sensory buffer). The additional module and connections (in dotted lines) involves a buffer to hold the corollary discharge of attention movement. This corollary discharge signal is employed to speed up movement of attention as well as prevent incorrect updating of the sensory buffer until attention has been moved to the correct place (as assessed by the corollary discharge buffer acting as an observer or attention state estimator).

Such identification of the corollary discharge of the signal of attention movement as equal to the (brief) signal of ownership of content requires much further evidence for support. One possibility is the dramatic state of PCE (Austin, 1998; Taylor, 2002b). This state, one in which there is no content of awareness (and to be identified with the ultimate meditatory states of nirvana or samadhi), would be created, in this model, by development through meditation of the ability to direct one's attention solely to the corollary discharge signal itself. All content, arriving from earlier sensory cortices, would be inhibited from gaining access to its working memory buffer. In such a state one would thus be attending solely to one's own attention movement. This has been suggested in the past (although unsuccessfully – Zahavi and Parnas,

1998) as a mechanism for the creation of consciousness, although not based on attention. Such a process is here given, for the first time, a specific neural circuitry to enable it to be achieved in the brain without the difficulties of a putative infinite regress (Shoemaker, 1968). The key to this is to use the mechanism of corollary discharge, in which there is a "splitting of the attention beam", so that one component can attend to its own copy. This process in particular provides specific predictions of the time course of neural activity in the appropriate sites, leading to predictions which can be used to test the hypothesis. This process of "inner attention" also needs to have an associated goal state set up in the frontal rules module. Frontal activity has indeed been observed during such meditation, so supports the approach proposed here (Taylor, 2002b).

A more cortically based approach can be developed, in which the apparent duplication of representations in the model (identity of the information content of the corollary discharge module acting as a "sentry at the gate", and the buffered working memory module which the sentry is guarding) gains a natural explanation (using the Felleman and Van Essen, 1991 connectivity model): in a multi-layer cortex the buffer for the efference copy of attention control is placed in layers 2/3, while the main buffer for the plant input is in the related layer 4. Such a placement makes more specific how the temporal dynamics of these layers of cortex should develop as consciousness of an input occurs (the activity in a particular layer 4 region, representing an item about to enter consciousness, being maximum after that of the associated region in layers 2/3). As such the model has testable local predictions, as well as predictions at a more global scale that could be detected by suitably sensitive brain imaging systems in studies of defects such as neglect.

8.9   Principles of the Global Brain

I have considered in detail a particular experiment (Schmitz et al., 1998; Taylor et al., 1998), as well as other evidence which supports the postulate that consciousness emerges from the joint activation of a network of areas. The details of the manner in which this occurs is at the heart of understanding the source of the extended duration of neural activity in the posterior slave working memory buffers in highest order multi-modal cortex. One explanation is in terms of "bubbles" of neural activity persisting in these cortical sites (Taylor, 1998), as I mentioned in the previous section.

I now want to try to bring these various features together to delineate several brain principles, which also extend discussion to the global brain.

1.   There exists evolutionary guidance for the brain's overall wiring diagram.

2.   There are several distinct control systems of cortical activity:

   • the TH/NRT/cortical gate (the "Conscious I");

- the TH/basal ganglia/cortical gate (the ACTION network);

- the limbic/motor interface (the MOVE network);

- the ARAS/brain stem core control system (the REFLEX network).

3. There exist specialized sites of long-duration activity (working memory sites).

4. There exists specialized long-term memory (LTM) to act as a highest level binding system and to create episodic memory and so self.

5. There exists binding/competitive processes (as under 2) for the interaction of

   - anterior cortex ↔ posterior cortex

   - anterior cortex ↔ limbic system

6. Consciousness is finally and lastingly created in a set of the specialized WM sites, in a dissociable manner, after suitable competition between cortical and memorial sites, as in the Relational Mind model (Taylor, 1999). A specific neural subsystem of the attentional control system, the CODAM structure, is crucial to grant the system the "ownership" component. Without this there is no sense of "what it is like to be".

Let me add some comments on these various principles. To begin, the above principles relate to the approach of Marr mentioned in the first section in terms of the implementation of specific algorithmic principles. The "Conscious I" TH/NRT cortical network is proposed to carry out a global competition between activities across large cortical distances. This is carrying out Marr's first level computational goal of giving a unified conscious experience as well as allowing a unified action response to inputs. It can also be implemented at an algorithmic level (Alavi and Taylor, 1995). The ACTION network is similarly carrying out the process of sequence learning and allowing actions to be developed appropriate to goals also held as part of the networks of Alexander et al. (1986). The algorithms involved can also be analyzed by simulation (Taylor and Taylor, 1998) and mathematics (Taylor and Taylor, 2000). The limbic-motor interface and the ascending reticular activating system (ARAS) and associated brain stem systems are still to be analyzed although their importance in creating primitive emotional states has been recently emphasized by Panskepp (1998). The existence of specialized sites of long-term activity is now extremely well documented, and the dynamics of these sites is being probed in ever increasing detail. The same can be said of the hippocampus, which is at the root of the special creation of long-term memory (and its subsequent laying down permanently elsewhere, although with special tags to hippocampus). These are thus wetware implementations (at Marr's level three) of his first level temporally extended traces and of long-term modifications. The algorithms to achieve these first level goals arc beginning to be put into place, such as in terms of recurrence of activity in specialized sites (the multi-modal sites of Mesulam, 1998) and convergence zones (Damasio, 1994; Moll and Miikulainen, 1997; Fellenz and Taylor, 1998). The binding/competition between anterior and posterior sites is only now emerging from brain imaging data, especially on tasks with increasing load. The nature of the algorithms to achieve the first level distribution of load between posterior (for encoding) and anterior (for active processing) is still to be determined.

Consciousness, as part of all this, is claimed to arise from suitably long-duration activity dominating other working memory activity, and thereby being able to send its message through the brain for general report. This is the special aspect of the CODAM subsystem, enabling overall control of attentional direction at a fast speed and so enabling rapid monitoring and response to environmental changes. Such a function is clearly satisfying the important task, for a large brain, of giving information rapidly to many distributed brain sites about salient processing results. From the present point of view it is clear that consciousness plays an important role of global filtering and communication as well as allowing fusion of conscious contents so as to store salient inputs in a manner allowing for effective retrieval from a large and complex data-base. If the CODAM model is correct then the crucial bridge between mind and matter will have been uncovered.

8.10   The Thinking Brain

Before concluding this rapid tour of the global brain, I would like to discuss that feature of the brain which gives us our unique ability to think, to reason, and to do so in language. What is it in the brain, and especially the cortex, which allows us to have developed these amazingly powerful modes of existence? I suggest it is the ability to store and manipulate sequences of neural activities, be they inputs from outside or internally generated "thoughts". There is also the ability to hold such activity over periods of time (many seconds) so they can be called on later for further use. The holding of activity has already been considered in terms of the ACTION network and of the related dedicated posterior working memory buffer sites. However, the abilities to perform temporal sequence storage and generation (TSSG) and to manipulate such sequences go considerably beyond these temporal features. Without such further powers we would be unable to manipulate any thoughts in our minds; they could only be experienced in a passive manner. With these powers we can move ideas around so that they achieve goals which have already been stored.

I earlier described some of the neural circuitry involved in action processing, that concerned with the frontal lobes. In particular the ACTION network and similar types of model has recently been applied to explore the processes underlying these higher order processes, considered as based on TSSG. Coupled nets that can perform TSSG have been developed by various groups. More specifically the ACTION network architecture has been applied to this problem (Taylor and Taylor, 1998, 2000; see Chapter 9 by Taylor and Taylor in this volume), as mentioned earlier. The relevant coupled modules have been shown to be able to give a basis for "chunking", or compression of signals in a temporal sequence, in a very specific form. Let me consider some of the problems that are faced when attempting such chunking, before turning to consider the importance of chunking for language and thought.

In a sequence of patterns $P_1, P_2,...$ the initial stage of chunking involves the creation of nodes encoding the pairwise transitions $P_1 \rightarrow P_2, P_2 \rightarrow ....$ Furthermore, there is experimental evidence that nodes exist in motor areas and in the basal ganglia encoding for triplet transitions $P_1 \rightarrow P_2 \rightarrow P_3$ and so on. Such encodings lead to an exponential explosion, in which it would be needed to have 2^n nodes to encode a sequence of length n. How can this explosion be avoided?

The simplest answer is to prevent it from ever being needed, by breaking sequences down into smaller lengths, say of length 3 or 4. Longer sequences are then built up by chunking the shorter sequences together. This seems to be what happens in our own brains, as noted in Greenfield (1991). There, Greenfield put forward evidence that as infants develop, their initial powers of object manipulation and of language are at the same level of ability, with sequences of length initially of one and then two stepping up to length three. The relevant neural systems were suggested as being identical. There then occurs a division of powers in which word sequences and object manipulation sequences are encoded in separate regions of BA 44/45. These control structures develop more anteriorly into BA 46 and BA 9 for language and object manipulation control respectively. The final story about how this is achieved is not known, but could be that the shorter temporal sequences are initially encoded in the more posterior portions of the prefrontal cortex and then higher chunking is achieved as more cortex comes available through myelination as the child develops. Thus the chunking that occurred would be of sequences of sequences of ..., up to a level of chunked sequences corresponding to the highest cortical level available to the human prefrontal cortex. In the process the melding together of syntax and of grammar in language would be occurring as part of the higher order coding process.

This "broad-brush" explanation of language encoding allows for a new perspective on the "deep structures" of Chomsky (1957). His suggestion of the existence of such structures deep in the brain is natural from this perspective, and even allows a new perspective on semantics. I would like to propose that semantics be equated with "virtual actions" in the following manner. The encoded sequence structures of manipulation and speech, developed as part of the prefrontal TTSG by the chunking process suggested above, are activated by the input of a word. This activation (now well recorded as part of brain imaging when semantic processing is occurring in a subject) does not produce the corresponding sequences of actions that occurred when the word was first met. However, such lack of actions does not mean that the neural structures available to make such actions are not partially available. It is that they are not fully activated, and that only a set of "virtual actions" occurs. The neural sites of such activity can be proposed as in basal ganglia and the associated cortical regions. The cortex is seen here to play a secondary role in the production of "virtual action"-based semantics.

The processes of manipulation of sequences, seen as part of planning and reasoning, can then be envisaged as occurring in the frontal lobes by means of the transformation of one cortical activity by another from a different module at a higher level. Such transformations are the life-blood of thinking and reasoning. They do not seem to be able to be performed by posterior cortical sites. Indeed this supports the epithet "passive consciousness" given earlier for that component of consciousness created by the posterior cortex. It is only the presence of an ACTION network kind of structure that would appear essential for such transformations to be achieved.

Such internal voluntary transformations would not appear to be able to be achieved until after several years of age, even as late as the beginning of the "concrete operations" stage of Piaget (at about 7 years of age). It is necessary for the developing child to create a suitable set of representations on which higher level control structures can act to transform between the lower level components. The internalisation of actions could be related to the development of internal semantics, since the internal actions are similarly not to be externalized. However, this would be at a later stage in child development that semantics first occurs, since it involves actions to be taken on the semanticized representations.

8.11   Discussion

The chapter started by discussing the problem of formulating principles of brain processing. It went on to consider brain imaging and some aspects of the associated data analysis. A description was then given of experimental results relevant to the search for consciousness in the brain. In particular, results obtained from global brain activity observed by fMRI during the Motion After-Effect were used to lead into the three-stage model of brain activity and consciousness. This model divides cortical (and associated sub-cortical) regions into those involved only with preprocessing (up to semantic level), those creating awareness at the lowest level through attention, and those involved with higher level control and attentional processing and the sense of self. This subdivision was also supported by further fMRI data, and is also strongly supported by data coming from patients suffering brain damage of a variety of sorts. It was given more detail by the CODAM model of creation of the pre-reflective self, itself a subcomponent of the overall attention processing system.

The three-stage model is only the beginning stage of the process of constructing more complex models allowing analysis of more complex activities involved in perceiving and reacting in numerous situations. Thus the development of object percepts and of episodic memory is needed to be modeled in order to allow the inner content of consciousness to be filled in, as well as to allow more detailed predictions to be made of brain activities expected to be observed during the solution of a range of tasks. Such a program has already been begun, and must go hand in hand with that of brain imaging itself in order to make sense of the wealth of such imaging data in respect of our own miraculous inner experience of consciousness.

In terms of some of the brain sites and their associated activities needed to create consciousness, various principles of the brain were developed in Section 8.7. These were then looked at further in the light of Marr's three levels of analysis (the computational theory, the algorithm used to solve the computation, and the hardware/wetware used to implement the computation). From the viewpoint developed in Section 8.7 it was realized that the three levels of approach were usefully intermixed so as to enable better understanding of the highest level computation from the lowest level wetware implementation. A more complete analysis of the increasing numbers of structural models of brain imaging data, across and increasing number of paradigms, will be expected to lead to greater clarity of the functions being performed by the separate modules of the networks, and thence begin to test these or similar principles being proposed for the whole brain. From brain imaging data the process is one of working from Marr's third level back up to the first. Finally, some brief ideas about cortical realizations of language, semantics and reasoning were given.

8.12   Acknowledgement

This chapter information was presented at the Workshop on Cortical Function, La Jolla, July 14–16, 1998, suitably updated to take account of more recent developments (as of February 10, 2002). The author would like to thank R. Hecht-Nielsen and J. Davis for their support to attend this meeting, and the ONR for funding.

References

Alavi, F., Taylor, J.G. (1995) A global competitive neural network. Biol. Cybernetics 72: 233–248.

Alexander, G.E., DeLong, M.R., Strick, P.L. (1986) Parallel organization of functionally segregated circuits linking basal ganglia and cortex. Annual Reviews of Neuroscience 9: 357–381.

Austin, J.H. (1998) Zen and the Brain. Cambridge, MA: MIT Press.

Baars, B. (1988) A Cognitive Theory of Consciousness. Cambridge: Cambridge University Press.

Baddeley, A. (1986) Working Memory. Oxford: Oxford University Press.

Bapi, R., Bugmann, G., Levine, D., Taylor, J.G. (1998) Analysing the Executive Function of the Prefrontal System: Towards a Network Theory. Behaviour and Brain Sciences (submitted).

Cabeza, R., Nyburg, L. (1997) Imaging cognition. J. Cog. Neuroscience 9: 1–26.

Chomsky, N. (1957) Syntactic Structure. New York: Mouton Press.

Crick, F.H.C. (1994) The Astonishing Hypothesis. Cambridge: Cambridge University Press.

Crick, F., Koch, C. (1998a) Consciousness and neuroscience. Cerebral Cortex 8: 97–107.

Crick, F., Koch, C. (1998b) Constraints on cortical and thalamic projections: The no-strong-loops hypothesis. Nature 491: 245–250.

Damasio, A.R. (1994) Descarte's Error. New York: Putnam.

Dennett, D.C. (1991) Consciousness Explained. New York: Little Brown.

Felleman, D.J., Van Essen, D.C. (1991) Distributed hierarchical processing in the primate cerebral cortex. Cereb. Cortex 1: 1–47.

Fellenz W., Taylor J.G. (1998) Enhancing the storage capacity of linear associative memories (in preparation).

Freeman, W. (1998) Contribution to the La Jolla workshop.

Greenfield, P. (1991) Language, tools and brain: The ontogeny and phylogeny of hierarchically organised sequential behaviour. Behavioural and Brain Sciences. 14: 531–595.

Halgren, E. (1994) Physiological integration of the Declarative Memory System. In: J. Delacour (Ed.) The Memory System of the Brain. Singapore: World Scientific, pp. 69–152.

He, S., Cavanagh, P., Intrilligator, J. (1996) Attentional resultion and the locus of visual awareness. Nature 383: 334–337.

Hecht-Nielsen, R. (1998) Contribution to the La Jolla workshop.

Karhunen, J., Oja, E., Wang, L., Vigario, R., Joutsenalo, J. (1997) A class of neural networks for independent component analysis. IEEE Transactions on Neural Networks 8(3).

Markram, H. (1998) Contribution to the La Jolla workshop.

Marr, D. (1982) Vision. New York: W.H. Freeman & Co.

Mesulam, M.M. (1998) From sensation to cognition. Brain 121: 1013–1052.

Moll, M., Miikulainen, R. (1997) Convergence-zone episodic memory: Analysis and simulations. Neural Networks 10: 1017–1036.

Nagel, T. (1974) What is it like to be a bat? Philosophical Review 83: 435–450.

Panskepp, J. (1998) Affective Neuroscience. Oxford: Oxford University Press.

Pollen, D. (1999) On the neural correlates of visual perception. Cerebral Cortex 9: 4-19.

Posner, M., Raichle, M.E. (1994) Images of Mind. New York: Scientific American Library.

Schmitz, N., Taylor, J.G., Shah, N.J., Ziemons, K., Gruber, O., Grosse-Ruyken, M.L., Mueller-Gaertner, H.-W. (1998) The search for awareness by the motion after-effect. Proceedings of the Conference on Human Brain Mapping 1998, NeuroImage.

Shoemaker, S. (1968) Self-reference and self awareness. Journal of Philosophy 65: 556–579.

Smith, E.E., Jonides, J., Marshuetz, C., Koeppe, R.A. (1998) Components of verbal working memory: Evidence from neuroimaging. Proc. Natl. Acad. Sci. (USA) 95: 876–882.

Taylor, J.G. (1995) Modelling the mind by PSYCHE. In: F. Soulie, P. Gallinari (Eds.) Proceedings International Conference on Artificial Neural Networks. Paris: EC2 & Co.

Taylor, J.G. (1998) Cortical activity and the explanatory gap. Consciousness and Cognition 7: 109–148 and 216–237.

Taylor, J.G. (1999) The Race for Consciousness. Cambridge, MA: MIT Press.

Taylor, J.G. (2000) The central representation: The where, what and how of consciousness. In: K.E. White (Ed.) The Emergence of Mind: Proc. Int. Symposium. Milan: Carlo Erba Fondazione, pp. 149–170.

Taylor, J.G. (2001) The importance of the parietal lobes for consciousness. Consciousness and Cognition 10: 379–417 and 421–424.

Taylor, J.G. (2002a) Consciousness: Neural Models of. In M. Arbib (Ed.) Handbook of Brain Theory and Neural Computation. Cambridge, MA: MIT Press.

Taylor, J.G. (2002b) Paying attention to consciousness. Trends in Cognitive Sciences 6: 206–210.

Taylor, J.G., Jaencke, L., Schmitz, N., Himmelbach, M., Mueller-Gaertner, H.-W. (1998) The three-stage model for awareness: Formulation and experimental support. NeuroReport 9: 1789–1792.

Taylor, J.G., Rogers, M. (2002) Simulating the movement of attention. Neural Networks (in press).

Taylor, J.G., Schmitz, N., Ziemons, K., Grosse-Ruyken, M.-L., Gruber, O., Mueller-Gaertner, H.-W., Shah, N.J. (2000a) The network of brain areas involved in the motion after-effect. NeuroImage 11: 257–270.

Taylor, J.G., Krause, B., Shah, N.J., Horwitz, B., Mueller-Gaertner H.,-W., (2000b) On the relation between brain images and brain neural networks. Human Brain Mapping 9: 165–182.

Taylor, J.G., Schmitz, N., Fellenz, W. (in preparation).

Taylor, N.R., Taylor, J.G. (1998) Experimenting with models of the frontal lobes. In: D. Heinke, G.W. Humphreys, A. Olson (Eds.) NCPW5 Connectionist Models in Cognitive Neuroscience, Perspectives in Neural Computation series. London: Springer, pp. 92–101.

Taylor, J.G., Taylor, N.R. (2000) Analysis of recurrent cortico-basal ganglia-thalamic loops for working memory. Biological Cybernetics 82: 415–432.

Tootell, R.B., Reppas, J.B., Dale, A.M., Look, R.B., Sereno, M.I., Malach, R., Brady, T.J., Rosen, B.R. (1995) Visual motion after-effect in human cortical area MT revealed by functional magnetic resonance imaging. Nature 375: 139–141.

Ungerleider, L.G., Courtney, S.M., Haxby, J.V.A. (1997) Neural system for human visual working memory. Proceedings of the National Academy of Sciences (USA) 95: 883–890.

Zahavi, D., Parnas, R. (1998) Phenomenal consciousness and self-awareness: A phenomenological critique of representational theory. J Consciousness Studies 5: 687–705.

Chapter 9

The Neural Networks for Language in the Brain: Creating LAD

N.R. Taylor and J.G. Taylor

9.1 Abstract

The neural networks that achieve linguistic skills in the brain are presently being uncovered by brain imaging methods using suitable psychophysical paradigms. We use these and other related results to guide the development of an overall neural architecture to implement Chomsky's "Language Acquisition Device" or LAD. We then consider in more detail the twin problems of the generation of infinite length sequences and the complexity of the recurrent system that produces such sequences. A recurrent neural network approach is used, based on our cartoon version of the frontal lobes, to analyze these two problems. The first is shown to be soluble in principle for any set of words by means of a set of "phrase analyzers", which contain complex neurones able to chunk suitable sequences. Further guidance from action and precept representations is indicated as helpful. The second problem is found to be solved by using the simplest level of chunking; this arises naturally in the learning process, according to a set of simulations, provided the task of language learning is suitably hard. We conclude with an overview of future developments to allow a full LAD to be developed so as to begin to approach adult speech.

9.2 Introduction

Language is arguably the most important feature of human intelligence. It grants us the ability to remove ourselves from the ongoing activities in the world and abstract our thoughts so as to solve difficult problems. Through such abstraction we can create concepts that allow us to understand the world. Natural language processing involves several areas of active research:

- formal linguistic analysis as seen through a logic framework;

- experimental observations and analyses of language in adults;

- language development in children;

- brain imaging;

- modeling of language processing.

Each of the areas has important contributions to make in the understanding of natural language processing (NLP), and all have active ongoing developments. An important recent approach in formal linguistics uses the temporal sequential order of word acquisition to build an increasingly well-defined syntactic tree as a sequence of words is heard or read (Pinker, 1995). The infinite generativity of language also provides important clues as to what infant language learning has to achieve (Gleitman and Leiberman, 1995). Experimental knowledge of how infants acquire language has been advancing considerably in terms of various semantic and syntactic frameworks (Thomas et al., 2001). Brain imaging is exposing the detailed networks involved in various linguistic tasks (Fiez, 1997; Gabrieli et al., 1998). Finally. modeling of language processing is a very active area of research, as the recent collections of articles indicate (Christiansen et al., 1999; Plunkett, 1998).

The problem of relating these various areas together is of great importance so that they can call on each others' strengths and insights, and also apply the constraints coming from one that can help guide development of another approach. The work presented in this chapter is to be seen as lying mainly under the heading of neural network modeling, but in the process calling on and relating to each of the other areas noted at the beginning. Thus it relates to the latest brain imaging results on natural language processing to guide the development of a multi-modular architecture able to learn to produce syntax and to recognize it in input word sequences. It also relates to the child language development field since we will only consider the early stages of syntactic learning, and with only limited sentence or phrase length. Finally it will also be concerned with the logical framework associated with the linguistic processing powers of the overall architecture being developed, in terms, say, of the framework of Kempson et al. (2000).

The basic problem we face in any approach to understanding language processing in the brain is: what is the prior structure we assume present? It is clear that children have a remarkable capacity to learn the complex rules of language, with seemingly far too little help from their environment. To explain this apparently miraculous language learning, Chomsky (1972) proposed that there is an innate internal system able to implement the rules of universal grammar. This device he termed a "Language Acquisition Device" or LAD. It was supposed to help solve the difficulty of learning language with no prior guidance. Speech heard by the child is processed automatically by LAD, thereby allowing the child to acquire the rules of language. The existence of such an a priori complex structure has been debated heatedly since then, with no final conclusion being reached. Much work has concentrated on the ordering of language acquisition, especially in comparison across languages. However, the debate has widened to include the acquiring of cognitive powers. Like the language abilities of young children, these also appear as if "de novo". Yet these powers also develop in a clear progressive manner (Goswami, 1998; Bowerman and Levinson, 2001). As suggested by a number of researchers (Haegman and Gueron, 1999), cognitive development is a crucial component of linguistic development, as well as having a separate non-linguistic aspect in its own right.

We emphasize in this chapter such a broader approach to language, with the inclusion of action processes at a cognitive level, following our earlier work (Taylor et al., 2000). This is achieved by using an action basis for the development of linguistic abilities. It is well known that the parts of the brain crucial to syntactic analysis are based in the frontal lobes, which are also crucial in action processing and planning. We therefore use common processing styles (sequence learning and generation) for developing both action and syntactic powers. At the same time we also consider the neural representations of objects that these actions are taken on to allow the development of action schemas. Meaning is given by actions taken on objects, so can be grounded in the world (so resolving the symbol-grounding problem). In this way the neural processing mechanisms used by LAD are only part of the overall processing powers of the frontal lobes, regarded as a system of modules able to perform transformations on object representations in posterior sites in the brain. To us, language is initially based on a symbolic description of such action processes. It then becomes elevated to an ever more abstract level by the development of semantic structures in other prefrontal as well as temporal lobe sites. The innate versus adaptive argument then is transferred from purely language learning to that of the whole gamut of cognitive powers in the child brain. We take a suitable fusion of prior genetically coded neural circuits with later learning through adaptation to the environment. The initial structure of LAD is that of the hard-wired early infant brain, with suitable genetic developmental processes allowing for later learning of higher-order structures as these further structures become on-line.

In more technical detail, the main thrust of the neural network approach we are following is initially to explore the neural powers of chunking and temporal sequence storage and generation (TSSG) achieved in a simplified model of the frontal lobes – the so-called ACTION network (Taylor, 1995) – in the development of NLP. It is guided by attempting to determine the answers to the following two questions:

Q1: How does syntax get bootstrapped from actions and sensory representations of objects at the early (and even possibly later) stages of language learning so as to become infinitely generative?

Q2: Does NLP require robustness in language learning, so that only the simplest chunking process can occur, with only two items chunked at a time?

With respect to Q1, infinite generativity appears at the second stage of language learning, after the initial one- and then two-word utterances are achieved. The manner these first stages are arrived at is unclear, but we suggest that prelinguistic concepts appear to be coded in children in close relation to the actions that can be made on the associated objects: actions speak louder than words initially in the developing infant. Later these action-based representations are activated on seeing an object, and so are expected to be of some relevance in word learning. From there to the development of words (verbs) for actions is a natural step; the learning of words (nouns) for objects will be expected also to associate the action-based representations. It is these latter which J.G. Taylor (1999a, 1999b) termed "virtual actions". Various approaches have also been developed in agent technology in which the interactions of agents with each other by simultaneously pointing to objects and naming them again indicates the value of an action-basis for language acquisition (Steels and Vogt, 1997; Hazlehurst and Hutchins, 1998). There has also been analysis of the relation between very early language acquisition and manipulative skills, and relation of these developments to parallels in frontal cortical development (Greenfield, 1991).

One feature to explore through neural simulation is the manner in which these action-based representations are able to guide the development of syntax at a low level; this has already been started by J.G. Taylor et al. (2000), where a simple architecture was developed that showed some value in the approach. They also initiated the study of neural modeling of this action-guided syntax learning. Our purpose here is to explore the further step that must be taken to achieve infinite generativity. This would need the recurrence of the neural ACTION network model of the frontal lobes. The problem we face relates to how such recurrence that is consistent with the neural architecture can be achieved. In particular, how does this lead to the ability to generate, say, strings $(A...AN)$ of arbitrary length for adjectival phrases (where A denotes an adjective and N a noun). What is the recurrent trick that achieves that? We will consider this, in general, in Section 9.4 and in more detail in Section 9.5, for the adjectival phrase analyzer in particular.

The second question is very apposite in the light of results on TSSG obtained by various neural modeling system (see the references in the following Section 9.3 for more details). As there seems to be no barrier to having considerable complexity in a temporal sequence, so the problem raised by Q2 is why very general syntactic language systems cannot be learnt, instead of those based on a restricted developing tree structure, with only two daughter branches per node (Greenfield, 1991). In contrast to this, the results of chunking through the ACTION network system has allowed chunking nodes of quite high complexity to be achieved. Even at the lowest non-trivial order, for sequences of length 3, the creation of, and need for, memory nodes which carry information about the whole sequence throughout its generation (and its reception once it has been learnt) would correspond to a tree structure with more than two daughter branches at each node. The basic phrase structure grammar (Steedman,

1999) does not correspond to that, but only to nodes with two daughters leaving them. Is this to allow more flexible linguistic powers than if higher complexity occurs? Or is there too great a complexity (and corresponding loss of robustness) in the creation of such a more complex syntactic structure? That seems to be the case, as far as robustness is concerned from the experience of TSSG by the ACTION net: higher order chunks are indeed harder to attain by learning, and use of levels of chunking have proved beneficial (Taylor and Taylor, King's College Technical Reports I–VII). We add parenthetically that the structural analysis of a story leads to more complex tree structures, with three or more daughters emanating from each node, so it may be that such higher order chunking does occur, but only for longer sequences stored in long-term memory. This question is considered in more detail in Section 9.6. This is only just a beginning of the construction of a neural network model of LAD. We discuss briefly in Section 9.7 further steps that need to be taken to continue that development. The chapter finishes with some conclusions.

9.3   The ACTION Net Model of TSSG

Models of temporal sequence storage and/or generation include: time delay neural networks (TDNN) (Waibel, 1989; Waibel et al., 1989), spin glasses (Hopfield, 1982), recurrent nets (Jordan, 1986; Elman, 1990) and recurrent temporal self-organizing maps. TDNNs are an application of Hakens embedding theorem. This states that measurements of a time series produced by an underlying dynamical system can always allow the underlying dynamics to be reconstructed by using an embedding vector of a time-lagged set of the elements of the measured series. The spin-glass approach of Hopfield (1982) is based on the creation of attractors in a recurrent net. The connection matrix provides for the transition from an attractor of one pattern of the sequence to one for the next. Probably the best known recurrent networks are those of Jordan (1986) and Elman (1990). Elman nets use context neurones to hold a copy of the previous activation levels of a hidden layer such that this history can be used to influence future outputs. Elman nets using Hebbian learning and leaky integrator neurones (Wang and Arbib, 1990a, 1990b) have been introduced into them which can learn and regenerate sequences involving repeated sub-sequences. The recurrent self-organizing map (RSOM) (Varsta et al., 1997) is an extension of the temporal Kohonen map (TKM) (Chappell and Taylor, 1993) that uses not only the history of previous inputs to find the best matching neurone for the current input pattern, but explicitly uses this history to adapt the weights. The RSOM, like the TKM, is a sequence classifier rather than a generator.

A number of problems occur with the above models. The spin-glass model of Hopfield was found by direct simulation not able to produce sequences greater than length four, and sequences of length four were not necessarily generated correctly (Taylor, 1998). The basic Elman net has problems with repeated input patterns and repeated sub-sequences, as well as when there is variation of the duration of inputs. The sequence generation models, of those mentioned above, produce sequences by transforming the current pattern into the next pattern of the sequence as an associative chaining system. They do involve chunking, allowing for sequences to be more efficiently encoded. However, they do not lead, as far as is known, to the specific class of chunking neurons observed in the monkey brain (Halsband et al., 1994). The models also lack topography, which occurs in many places in the brain. A further difficulty for these models is over the storage and generation of sequences with delays. To cope with delays the models need to have an input and output that represent delays, but a delay is a lack of an output for some period of time. Such a representation is not natural in these models. Finally it is not obvious how automatization of motor control would arise without having an architecture allowing direct interaction from posterior sites with the basal ganglia (hidden) nodes to develop as part of the learning process. It is possible that such recurrent neural models can achieve such a two-stage form of processing, but we have not seen that described. There is clear experimental observation of the siting of such automatization, and we believe that it needs a more detailed relation to brain architecture to be achieved, as we now describe.

Modification of recurrent networks to incorporate a simplified version of the architecture of the frontal lobes was achieved by the ACTION network of J.G. Taylor (1995, 1996), Monchi (Monchi, 1998; Monchi and Taylor, 1995, 1997, 1999), Taylor and Alavi (1996), and N.R. Taylor (Tech. Reports; Taylor and Taylor, 1998, 1999a, 1999b, 2000a, 2000b). We include here a brief account of this processing style. We consider first the nature of the frontal lobes themselves. There are no comparable subcortical structures accompanying the posterior cortex to those for frontal regions. The basic feature of the sub-cortical components of these frontal areas are the basal ganglia, two-layer systems activated unidirectionally from the cortex, the upper layer being the striatum (from its appearance), being decomposable into caudate and putamen. Beneath these, and close to the relevant part of the thalamus (the mediodorsal nucleus) is the globus pallidus, comprising two components, termed internal and external. The unidirectional flow of information is :

cortex → striatum (STR)→ globus pallidus (internal) (GPi)
→ thalamus (TH) → cortex

so there is a feedback of activity round the above "long" loop. There may also be similar feedback more rapidly round the "short" cortex ↔ TH loop, and although there is known topography in these connections the presence of any reverberation of such feedback activity is unknown. There is also the presence of the indirect loop, involving the globus pallidus external (GPe) as well as the sub-thalamic nucleus (STN), as shown in Figure 9.1, a schematized version of the frontal lobe architecture we have termed elsewhere the ACTION network, since it is so important for the generation of actions.

The ACTION network has been used to model TSSG (Taylor, 1995, 1996; Monchi, 1998; Monchi and Taylor, 1995, 1997, 1999; Taylor and Alavi, 1996; Taylor and Taylor, Tech Reports, 1998, 1999a, 1999b, 2000a, 2000b). In these papers it was shown how, either in suitably hard-wired architectures or in trained ones, there appear a set of neurones which achieve chunking of the temporal sequences, and which agree with the activity patterns of neurones observed by researchers in single cell recordings in monkeys. These neurones have continued activity during delay periods. We classified them as various sorts of initiator, transition and complex memory neurones. These types of neurones are sequence-specific; in addition pre-movement and movement neurones are recorded.

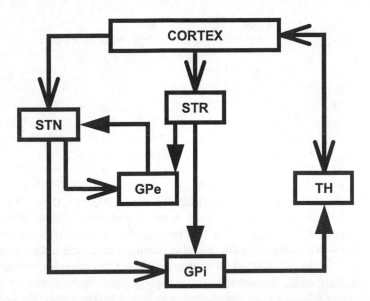

Figure 9.1. The ACTION network. Excitatory connections are indicated by open arrowheads and inhibitories by closed.

To summarize what has been shown in more detail using this architecture:

1. A suitably hard-wired version of the ACTION network can simulate the phasic and tonic cells recorded in monkey SMA during TSSG tasks with a manipulandum. The detailed temporal character of the cells observed is reproduced as part of the dynamics of the ACTION network, with clear differentiation of the tonic and phasic neurones as is their use in chunking the temporal sequence of movements involved;

2. Suitable modeling of DA reduction can be achieved by modification of the striatal output thresholds to the GPe and GPi as well as laterally. These are known to correspond to D2 and D1 receptors in the first and to the second and third cases respectively. Thus DA reduction can be modeled as a reduction in the first and an increase in the second and third of these threshold values. This causes a reduction of activity in GPe and increases in STN and GPi. Such modifications are observed in monkeys treated with

MPTP, which causes PD-like symptoms and destroys cells in the SNc. Modifications of TSSG for learnt sequences were also consistent with PD-like symptoms. The results were consistent with the PD model of DeLong, in which excess STN activity causes strong inhibition of TH, and thereby rigidity and reduction of movement;

3. In a converse manner DA increase was modeled, as occurs in Huntington's chorea, by opposite changes to the thresholds for striatal to GPe, GPi and laterally on STR compared to those made in (2) above. As to be expected, opposite changes occurred to the GPi, STN and GPe, with the first having decreased activity and little changes in the other two. It is the decreased inhibition from GPi, brought about by these parameter changes, which allows for increased spontaneous activity of TH and hence corresponding movements. It is these changes, it was suggested, which are related to the symptoms of a patient suffering from Huntington's chorea.

4. Finally the source of these changes was analyzed mathematically in terms of the underlying working memory ability of the ACTION network. This had been analyzed earlier (Taylor and Taylor, 2000a) in terms of a reverse saddle-node bifurcation of the fixed-point equations for the asymptotic cortical activity arising from the equations. The analysis showed how the coupled recurrent loops on longer sustained movements can fail. TSSG cannot be attained by the system any more after this change of the threshold parameters. In the Huntington's chorea case the system leads to excessive and spontaneous movement, as corresponds to one of the symptoms of the disease.

We develop in the next section the general framework of phrase analyzers, based on the chunking nodes we have recognized as occurring naturally in the ACTION network during learning of TSSG; we add to these the action/concept basis for guidance of the initial (and possibly later) stages of language learning.

9.4   Phrase Structure Analyzers

The development of syntax through the first 18 months of a child's life (Pinker, 1994, 1995) is that first, single words are learnt (by the end of the first year), and then secondly two-word sentences are produced. It is the thesis here that effective learning of syntax at this early stage is guided by the set of learnt actions on the objects for which nouns have been learnt, in association with the learnt actions for the (action) verbs. That this is a reasonable approach to the "bootstrapping" of syntax, by associating it with action-based semantics, has already been argued by one of us elsewhere (Taylor, 1999a; Taylor et al., 2000), on the basis of brain imaging results which detect strong cortical activations in prefrontal sites during semantic processing (Jennings et al., 1998; Fiez, 1997). There are also interesting features of conceptualization by infants in terms of combining objects they have played with by means of the similar actions they can take on them.

A possible database for attacking this problem will be based on the discussion in Pinker (1994, 1995). There it is stated that, by the end of the first year, about half the words are for objects: food, body parts, clothing, vehicles, toys, household items, animals and people. The others are words for actions, for motions, for routines, especially social ones, and modifiers. Later two-word strings are of the form: *All dry; more cereal; boot off; see pretty; papa away; see baby.* In order to be most effective we take a reduced database which has only concrete nouns and action verbs, but one which allows for the full case-grammar set of agent, action, recipient, object and location (following the table of two-word sentences on p. 143 of Pinker, 1994).

The words we take are:

> Nouns: *mommy, baby, car, teddy, ball, floor, door, table, doggie, cereal.*
> Verbs: *reach, grasp, release, stroke, push, shut, touch, lift, eat.*

According to our above description of how syntax arises from action-based semantics, sequences of actions that an infant has taken and learnt on the objects need to be described, so that these can be coupled to the appropriate actions for the verbs. In this way, for example, an infant that has pushed a car around, when told that the object it has in its hand is "car" and that it learns it is "pushing" that car, will then say "push car" (or "car push" according to the language parameters determining ordering) in which the action verb functions as a verb phrase, the object as the noun phrase in the incipient phrase structure analysis: VP = (push = V)(car = N). However, before the relation between objects, actions and their descriptor words is attempted we will develop the gluing together of the object representations and the actions with which they are associated.

Detailed analysis of the sequences of actions on the objects in terms of the relevant verbs is therefore developed next. Some of the possible actions that can be taken on each of the objects (we only consider a few representatives) are as follows:

> Single actions: *reach, grasp, lift, push, stroke, put in mouth.*
> Sequences of actions for objects:
> > car: reach hand→touch car→grab car→push car→release car;
> > teddy: reach hand→touch teddy→stroke teddy→release teddy;
> > cereal: reach hand→touch spoon→grab spoon→lift spoon→
> > spoon in mouth.

We turn now to develop a general neural model for language learning using visual representations of objects and representations of actions for verbs. For the former may be chosen a SOFM, since this allows a simple localized interpretation; the nodes of the SOFM are leaky integrators, so as to form a buffer working memory site, as is observed in parietal/temporal lobes. The object coding is by dedicated nodes, for simplicity.

There are two early stages of language learning that, we suggest, occurs in the developing infant:

1. learning action-based semantics for sensory concepts (even pre-linguistically);

2. learning to relate representations for words to those for concepts, including their action basis.

We consider these stages successively here.

The actions are coded in an ACTION network to allow for sequence learning (Taylor and Taylor, 2000a, 2000b). This architecture is based on the recurrent neural structures observed in the frontal lobes which allow for activity to pass through several loops: cortex→ cortex, cortex→basal ganglia→thalamus→cortex, in which the latter loop has several subvarieties. One of the important features discovered from work on monkeys (Halsband et al., 1994) is the existence of "chunking" nodes, which are created in the SMA to handle temporal sequence storage and generation (TSSG). These nodes have also been produced in simulations of the ACTION network as part of TSSG (Taylor and Taylor, 2000a, 2000b). It is these chunking nodes which will be of crucial use in the development of syntax later in the chapter. Each action of the set specified in the previous section is coded as a single cortical node in an ACTION net, with its associated basal ganglia/thalamic loops.

The action sequences associated with a given object representation are now linked to it by a Hebbian learning process, in which the relevant action nodes are activated (according to the actions being taken) by the particular active object representation. On future activation of this object representation the action representations will also be activated. However, the actions will not actually be taken, but only be primed, as shown in Figure 9.2, since no source of "willed intention" is present in this case (the subject does not wish to perform the actions, although it did so during the earlier learning period) to prime the ACTION network system. This is one way of achieving semantics as "virtual actions"; there may be others.

The virtual action basis for semantics is now extended to words by the introduction of two new ACTION networks, one for words for the actions and one for the objects introduced earlier. The appropriate connections between the relevant action (object) representations modules and the verb (noun) representation modules is achieved by further Hebbian learning of the bi-directional connections between the object representation and object word representation (noun) modules and between the action representation and action word representation (verb) modules.

In Figure 9.2 the willed intentions module is only active when an action is being taken; learning then takes place between the action representations and the associated object representations. Similar activity in the action representations during learning of the associated words to objects (or the actions themselves) leads to learning not only of the connections between object representations and associated noun representations (as shown by the bi-directional line joining the relevant nodes in each of these modules in Figure 9.2) but also of the line joining the noun representation and the set of relevant actions in the action representation module. Inactivation of the willed intentions module during later activation of the objects does not prevent the associated action representations from being primed by the object representations, so leading to guidance of the associated word representations by these "virtual actions". This will be used in the following section for learning the basic syntax of two-word phrases.

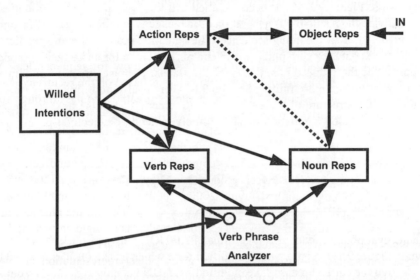

Figure 9.2. The single word modules and their action-based semantics. The verb phrase analyzer has two nodes specified: the one on the left is an initiator, and is activated by the willed intention to generate a verb (guided by the relevant action representation). This then activates in its turn the transition node (on the right), which activates a noun (also guided, now by the relevant object representation). The sequence then ends by suitably applied inhibition.

Recurrent nets have been used considerably to attack the problem of learning syntax for context-free grammars (Steedman, 1999; Ho and Chan, 1994). Having chosen to use actions as a basis of semantics in the previous section, and developed for it a specific "virtual actions" recurrent neural architecture, we will now turn to apply this to guidance for the development of syntax for two-word sentences.

The next stage of language production in the developing infant is indeed that of two-word sentence production. In English, sentences like "push car" or "stroke teddy" are produced, as are "baby cereal" or "baby push". The first two of these are verb phrases (VP), the second two are sentences, following the standard decompositions VP→VN, S→NP VP. To learn to produce these sequences of length 2, two new types of nodes are needed: initiator nodes (IN) and transition nodes (TRANS), so as to have the more complete decompositions:

$$IN(VP) \rightarrow V \rightarrow TRANS(V \rightarrow N) \rightarrow N$$
$$IN(S) \rightarrow NP \rightarrow TRANS(NP \rightarrow VP) \rightarrow VP$$

In these expressions IN(X) denotes the initiator nodes for the sequence X, TRANS(X→Y) the transition nodes from the component X to Y in the learnt and reproducible sequence (XY).

The learning of such transitions has been discussed elsewhere (Taylor and Taylor, 2000a, 2000b); chunking modules of a system able to produce such a transition are also shown in Figure 9.2. It has been assumed that the "willed intention" module is able to activate them suitably if generation of language is required or also if recognition is involved of an incoming sequence. The chunking module shown specifically in Figure 9.2 is only for verb phrases, and contain only initiator and transition nodes, as required for the reduced complexity of the standard phrase structure grammars; this will be discussed more fully in Section 9.6. There is also no infinite generativity contained in the structure of Figure 9.2; we will remedy that in the next section, by developing a detailed simulation of an infinitely generative adjectival phrase analyzer.

9.5   Generativity of the Adjectival Phrase Analyzer

The adjectival phrase model uses two forms of ACTION network, one with the form shown in Figure 9.1, the other with a modified form which has an OUT region fed by activity from the cortical region. The modified ACTION net only generates an output if the OUT neurone exceeds some threshold value. Figure 9.3 indicates the wiring diagram for this model.

The model is run by presenting phasic input to the feature representations (any number) as well as to the object representations (only one), generating sustained activity of the excited representations. The activity in the representations (features and objects) produces low-levels of activity in the adjectival representations and noun representations which inhibit those adjectives and nouns without excited features and objects via the lateral STR connections. Therefore we have priming of a number of adjectives and one noun. The initiator is excited by a phasic willed intention input. At the same time another willed intention causes inhibition to all weights from the adjectival representations to the transition neurone, this input can be linked to excited feature representations such that it is removed when there are no more features excited. The initiator excites all adjectival representations, whether they are primed or not, however only those that are primed by feature representations can produce sustained activity and excite their OUT neurone beyond the threshold value. The striatal lateral connections between adjectives also ensures that only one primed adjective exceeds the threshold at a time hence only one word can be generated at a time. Once an adjective exceeds the threshold that word is generated and inhibition turns off the feature and adjective associated with the word.

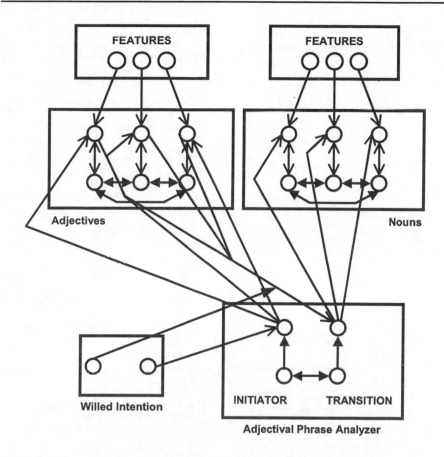

Figure 9.3. The wiring diagram of the adjectival phrase analyzer. The feature representations and object representations are composed of recurrent neurones. ACTION networks are indicated by linked vertical pairs of neurones, the upper neurone is cortical whilst the lower is situated in the STR. The adjective and noun representations are composed of modified ACTION nets, and the adjectival phrase analyzer is composed of un-modified ACTION nets. Excitatory connections are indicated by open arrowheads and inhibitories by closed arrowheads. Inhibition via proprioceptive feedback to all areas is not shown.

If the inhibitory willed intention is still active then the transition neurone is not excited and the initiator neurone remains active such that another adjective can be generated. As mentioned above, the willed intention that inhibits all inputs to the transition neurone can be linked to active feature representations if no more features are active then the inhibition is removed allowing excitation of the transition neurone by the last adjective. The transition neurones excites all noun representations but only the primed noun can exceed its threshold and generate the noun. Once the noun is generated the whole model is inhibited, turning off object and noun representations, the transition neurone, and any un-generated features and adjectives. Figure 9.4 shows the time-courses of some of the neurones involved in generating an adjectival phrase composed of three adjectives.

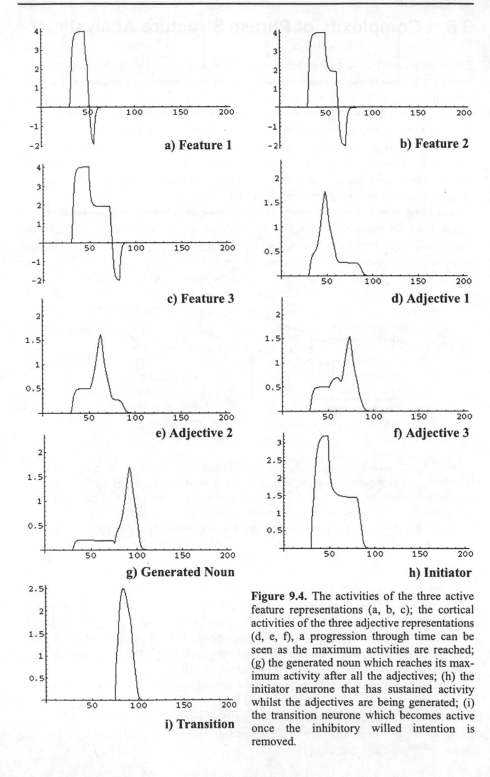

a) Feature 1

b) Feature 2

c) Feature 3

d) Adjective 1

e) Adjective 2

f) Adjective 3

g) Generated Noun

h) Initiator

i) Transition

Figure 9.4. The activities of the three active feature representations (a, b, c); the cortical activities of the three adjective representations (d, e, f), a progression through time can be seen as the maximum activities are reached; (g) the generated noun which reaches its maximum activity after all the adjectives; (h) the initiator neurone that has sustained activity whilst the adjectives are being generated; (i) the transition neurone which becomes active once the inhibitory willed intention is removed.

9.6   Complexity of Phrase Structure Analysis

The problem we face with the complexity of language has already been discussed in Section 9.2: why is the language recognition and production mechanism in the brain not more complex? That is essentially the content of Q2 of the introduction: why can we not produce chunking neurones at a higher complexity than length 2 for the temporal sequences of speech? That the content of natural speech is such as not to require such higher complexity is no argument, since the evolution of speech very likely went hand in hand with the evolution of the production apparatus in the brain. We must therefore look more closely at the nature of this mechanism, as a neural system, to understand the problem. This we do briefly here.

We have used a previous model that can learn to generate internal sequences of length 3 to investigate how noise affects the length of sequence that can be learnt and regenerated. The initial model, without noise, had clean inputs hence only one input is on at a time; the noisy version has an input that is on maximally and the other inputs are also excited to some level. We have chosen to add noise at the rates of 5, 10, 20, 30, 40 and 50 percent of maximal activity. Currently the success rate of the noisy models to learn and generate a length 3 sequence is 5 percent, the other 95 percent of simulations either generate an internal length 2 sequence or can only generate the first output correctly.

We conclude from this result that a noisy input makes it very difficult for the creation of the necessary chunking nodes to be able to learn and regenerate sequences of length more than two. In other words noise has caused degradation of the ability to learn anything more complex. This is a natural result: noise will always cause a certain degree of degradation of representations. In the temporal domain in particular noise compounds in time. Thus it will put a limit on the length of sequences which can be chunked. This is the basis of our explanation of the phrase structure of natural language.

9.7   Future Directions in the Construction of LAD

So far in this chapter we have considered the somewhat technical questions of how syntax could arise from neural structures known to be present in the developing infant brain, and what might be the general level of generativity of linguistic production. However, this only begins to scratch the surface of creating an effective LAD, able to develop speech comprehension and production up to adult competence. This is a non-trivial program to develop. We can only give here a broad survey of further steps in the overall LAD development program.

The steps themselves must consist of the construction of the following systems:

- A semantic system, able to allow the creation of the semantics of abstract concepts, so not to be directly related to actions in the frontal lobes.

- A working memory/attention control system, allowing the learning of new words and meanings, and to develop comprehension powers.

- A long-term memory system, allowing for past inputs and outputs to be stored at a suitably coded level so as to be used in discourse at a later time.

- A motivational reward system used in the learning process of the above systems.

These systems are all available as off-the-shelf neural network systems in some form or other. Thus semantic modules can be developed by use of the self-organizing learning rule (Kohonen, 1998). This enables higher-order categorization, for nouns as descriptors of objects in the world, by initial development of a semantic map at basic category level. This uses as initial input object feature maps and associated basic category nouns. Further modules are then brought on-stream to learn the names of higher-order categories. This uses the output of the basic category noun maps as well as the names of the higher-order categories (Stanley, 2001). Thus "dog" is first learnt, as well as other basic category words, as descriptors of basic category visual features of inputs. Then this learnt system is used to guide the learning of "terrier" for the appropriate more specific input features of a terrier. In this manner an initial "guided" approach is suggested for learning semantics.

The facility of working memory has been combined with that of attention, since there is good evidence that these involve modules that may be identical. There are several recent models of attention in the brain (Taylor, 2002); in an adjoining chapter in this book a discussion is given of its overall controlling power for achieving suitable responses. In LAD attention is used for single word learning, since it is known that infants learn words through attending jointly with their care-givers to objects in the environment (Silven, 2001). At the same time attention is used as part of the control system for language response, both with generation and monitoring of speech being used. Thus LAD has attention as an overall controller to determine what is said, what is understood and what is remembered.

Related to attention are long-term memory capabilities needed in the brain. The memorization process is very likely turned on by attention processes, so that what is being attended to (what is in awareness) is remembered. This puts an important constraint on LAD architecture, as well as indicating that in language comprehension there is need for the process of "automatization". The ability to change an attentionally driven motor process to one automatically produced is a crucial component of human skill development. Such processing involves the frontal lobes in interaction with the parietal sites as the source of attention control. Automatization allows attention to be shifted from the learnt motor process to other tasks of more pressing concern. Return of attention to a learnt motor act is known to cause loss of skill, in terms of "choking" of the skilled performer. There appears to be shifting the ongoing prefrontal cortical representation of the motor act to a basal ganglia-parietal form as automatization is achieved. This automatic run-off of sequences of motor acts is halted if attention is later alerted by any error arising in the sequencing. Attention then swings back to the sites from which the error arose so as to correct response. Such a process of automatization must be at work in speech comprehension. This can occur in LAD by means of the descent from cortical to striatal activity as syntactic and schemata analysis occurs in the ACTION network. Only higher-order activations would then arise to awareness/attention as phrases were processed to activate such representations. This would correspond to "gist" detection in a speech input. The same use of higher-order attended activations in the schemata regions of LAD's frontal lobes would also arise in speech production: LAD would only need to be aware of the general gist of what is wanted to be said at any time. The detailed form of that speech output would then arise automatically from pre-learnt speech phrases.

Finally there is the motivational component in speech learning and generation. That infants are motivated to learn speech is not in question. It is uncertain what is the underlying brain motivational system achieving that. In LAD we assume there is a motivational system based on reward learning, the rewards being developed from earlier interactions with the teacher at more primitive levels – for single objects recognition, etc. The reward system in LAD is ubiquitous, and is related to the interacting dopamine and related neurotransmitter systems (acetylcholine, serotonin, noradrenaline).

The steps for developing LAD with linguistic abilities up to those of a 6-year old infant have been laid out carefully (Taylor, 2001), following the careful descriptive work of Brown (1973). However, the manner in which this progressive learning will need related cognitive powers is yet to be determined; we hope to report on that elsewhere.

9.8   Conclusions

In this chapter we have presented a description of an overall neural approach to speech development: the neural system we call LAD. Preliminary answers are given to two technical questions for LAD: what is a natural neural architecture for the infinite generativity of speech, and what is the neural reason behind the simplicity of the phrase structure of natural language syntax. The answers are that suitable specific recurrent modules ("phrase analyzers") must be constructed to decompose or generate syntactically correct speech, whilst if the input is suitably degraded it will not be possible for the TSSG system to learn more than the simplest sequence structure, that of sequences of length 2.

The results presented above can be checked. Thus, if it were possible to perform single cell analysis on a speaking human, or one listening to speech, then the chunking modules – the phrase analyzers – should be detectable. That is not an easy experiment to perform, not least because of ethical considerations. However, use of MEG systems, for temporal specificity, in combination with fMRI for spatial accuracy, would allow a beginning of the analysis for spatial and temporal separation of the phrase analyzers we suggest are present.

In the previous section we then outlined very briefly other aspects of concern of LAD and how they might be tackled. We refer the interested reader to the website for further information (http://www.lobaltech.com).

References

Bowerman, M., Levinson, S.C. (Eds.) (2001) Language Acquisition and Conceptual Development. Cambridge: Cambridge University Press.

Brown, R. (1973) A First Language. Cambridge, MA: Harvard University Press.

Chappell, G.J., Taylor, J.G. (1993) The temporal Kohonen map. Neural Networks 6: 441–445.

Chomsky, N. (1972) Language and Mind. New York: Harcourt Brace.

Christiansen, M.H., Chater, N., Seidenberg, M.S. (Eds.) (1999) Connectionist models of human language processing: Progress and prospects. Special issue of Cognitive Science 23(4): 415–634.

Elman, J.L. (1990) Finding structure in time. Cognitive Science 14: 179–212.

Fiez, J.A. (1997) Phonology, semantics and the role of the left prefrontal cortex. Human Brain Mapping 5: 79–83.

Gabrieli, J.D.E., Poldrack, R.A., Desmond, J.E. (1998) The role of the left prefrontal cortex in language and memory. Proc. Natl. Acad. Sci. USA 95: 906–913.

Gleitman, L.R., Leiberman, M. (Eds.) (1995) Language. Cambridge, MA: MIT Press.

Goswami, U. (1998) Cognition in Children. Hove, East Sussex: Taylor & Francis.

Greenfield, P.M. (1991) Language, tools, and brain: The ontogeny and phylogeny of hierarchically organized sequential behavior. Behavioral and Brain Sciences 14: 531–595.

Haegman, L., Gueron, J. (1999) English Grammar. Oxford: Blackwell.

Halsband, U., Matsuzaka, Y., Tanji, J. (1994). Neuronal activity in the primate supplementary, pre-supplementary and premotor cortex during externally and internally instructed sequential movements. Neuroscience Research 20: 149–155.

Hazlehurst, B., Hutchins, E. (1998) The emergence of propositions from the co-ordination of talk and action in a shared world. Language and Cognitive Processes 13: 373–424.

Ho, E.K.S., Chan, L.W. (1994) How to design a connectionist holistic parser. Neural Computation 11: 1995–2016.

Hopfield, J.J. (1982) Neural networks and physical systems with emergent collective computational abilities. Proceedings of the National Academy of Sciences USA 79: 2554–2558.

Jennings, J.M., McIntosh, A.R., Kapur, S. (1998) Mapping neural interactivity onto regional activity: An analysis of semantic processing and response mode interactions. NeuroImage 7: 244–254.

Jordan, M.I. (1986) Attractor dynamics and parallelism in a connectionist sequential machine. Proceedings of the Cognitive Science Society (Amhurst, MA), pp. 531–546.

Kempson, R., Meyer-Viol, W., Gabbay, D. (2000) Dynamic Syntax. Oxford: Blackwell.

Kohonen, T. (1998) The Self-Organising Map. Berlin: Springer.

Monchi, O. (1998) Modelling Functions and Dysfunctions of the Anatomical Circuits Involved in Anterior Working Memory and Attentional Tasks. PhD Thesis, Department of Mathematics, King's College, University of London (unpublished).

Monchi, O., Taylor, J.G. (1995) A model of the prefrontal loop that includes the basal ganglia in solving the recency task. Proc. WCNN'95, Washington, DC, Vol. 3, pp. 48–51.

Monchi, O., Taylor, J.G. (1997) In: J.A. Bullinaria, D.W. Glasspool, G. Houghton (Eds.) 4th Neural Computation and Psychology Workshop: Connectionist Representations. London: Springer-Verlag, pp. 142–154.

Monchi, O., Taylor, J.G. (1999) A hard wired model of coupled frontal working memories for various tasks. Information Sciences, 113: 221–243.

Pinker, S. (1994) The Language Instinct. London: Penguin Press.

Pinker, S. (1995) Language acquisition. In: L.R. Gleitman, M. Leiberman (Eds.) Language, Vol. 1. Cambridge, MA: MIT Press, pp. 135–182.

Plunkett, K. (Ed.) (1998) Language Acquisition and Connectionism. London: Psychology Press.

Silven, M. (2001) Attention in very young infants predicts learning of first words. Infant Behaviour and Development 24: 229–237.

Stanley, P. (2001) Aspects of Language Modelling by Neural Networks. King's College M.Sc., Information Processing and Neural Networks Thesis, University of London (unpublished).

Steedman, M. (1999) Connectionist sentence processing in perspective. Cognitive Science 23: 415–634.

Steels, L., Vogt, P. (1997) Grounding adaptive language games in robotic agents. In: I. Harvey, P. Husbands (Eds.) Proceedings of the 4th European Conference on Artificial Life. Cambridge, MA: MIT Press.

Taylor J.G. (1995) Modelling the mind by PSYCHE. In: F. Fogelman-Soulie, P. Gallinari (Eds.). Proc. ICANN'95. Paris: EC2 & Co, pp. 543–548.

Taylor, J.G. (1996) New models of control from biology. In: I. Parmee, M.J. Denham (Eds.) Adaptive Computing in Engineering Design and Control 1996

(ACEDC'96), Proceedings of the 2nd International Conference of the Integration of Genetic Algorithms and Neural Network Computing and Related Adaptive Techniques with Current Engineering Practice.

Taylor, J.G. (1999a) Do virtual actions avoid the chinese room? In: J. Preston, M. Bishop (Eds.) Views into the Chinese Room: New Essays on Searle and Artificial Intelligence. Oxford: Clarendon Press.

Taylor, J.G. (1999b) The Race for Consciousness. Cambridge, MA: MIT Press.

Taylor, J.G. (2001) LAD: The Long Haul. Lobal Technologies Internal Report. (http://www.lobaltech.com).

Taylor, J.G. (2002) Paying attention to consciousness. Trends in Cognitive Sciences 6: 206–210.

Taylor, J.G., Alavi, F.N. (1996) A basis for long-range inhibition across cortex. In: J. Sirosh, R. Miikulainen, Y. Choe (Eds.) Lateral Interactions in Cortex: Structure and Function. http://www.cs.utexas.edu/users/nn/web-pubs/htmlbook96/.

Taylor, J.G., Taylor, N.R., (2000a) Analysis of recurrent cortico-basal ganglia-thalamic loops for working memory. Biological Cybernetics 82: 415–432.

Taylor, J.G., Taylor, N.R., King's College Technical Reports: BT Reports I–VII. London.

Taylor, J.G., Taylor, N.R., Apolloni, B., Orovas, C. (2000) Constructing symbols as manipulable structures by recurrent networks. In: Proceedings of IJCNN'2000, Vol. 2.

Taylor, N.R., (1998) Temporal Sequence Storage by Neural Networks. PhD Thesis, Department of Mathematics, King's College, University of London.

Taylor, N.R., Taylor, J.G. (1998) In: D. Heinke, G.W. Humphries, A. Olsen (Eds.) Connectionist Models in Cognitive Neuroscience: The 5th Neural Computation and Psychology Workshop. London: Springer-Verlag, pp. 92–101.

Taylor, N.R., Taylor, J.G. (1999a) Learning to generate temporal sequences by models of frontal lobes. IJCNN'99 Proceedings. Erlbaum.

Taylor, N.R., Taylor, J.G. (1999b) Modelling the frontal lobes in health and disease. ICANN'99 Proceedings. IEEE Press.

Taylor, N.R., Taylor, J.G. (2000b) Hard-wired models of working memory and temporal sequence storage and generation. Neural Networks 13: 201–224.

Thomas, M.S.C., Grant, J., Barham, Z., Gsödl, M., Laing, E., Lakusta, L., Tyler, L. K., Grice, S., Paterson, S., Karmiloff-Smith, A. (2001). Past tense formation in Williams syndrome. Language and Cognitive Processes: 16 (2/3), 143–176.

Varsta, M., Millan, J. del R., Heikkonen, J. (1997) A recurrent self-organising map for temporal sequence processing. Proceedings of ICANN'97, pp. 421–426.

Waibel, A. (1989) Modular construction of time-delay neural networks for speech recognition. Neural Computation 1: 39–46.

Waibel, A., Hanazawa, T., Hinton, G., Shikano, K., Lang, K. (1989) Phoneme recognition using time-delay neural networks. IEEE Transactions on Acoustics, Speech and Signal Processing 37: 328–339.

Wang, D.L., Arbib, M.A. (1990a) Complex temporal sequence learning based on short-term memory. Proceedings of the IEEE 78(9): 1536–1543.

Wang, D.L., Arbib, M.A. (1990b) Timing and chunking in processing temporal order. IEEE Transactions on Systems, Man, and Cybernetics 23: 993–1009.

Chapter 10

Cortical Belief Networks

Richard S. Zemel

10.1 Abstract

Most theoretical and empirical studies of cortical population codes make the assumption that underlying neuronal activities is a unique and unambiguous value of an encoded quantity. We propose an alternative hypothesis, that neural populations represent, and effectively compute, probabilities. Under this hypothesis, population activities can contain additional information about such things as multiple values of, or uncertainty about, the quantity. We discuss methods for recovering this extra information, and show how this approach bears on psychophysical and neurophysiological studies. A natural extension of this probabilistic interpretation hypothesis casts interacting populations as a belief network, a structure which permits the analysis of information propagation from one population to another. This novel framework for population codes opens up new avenues for studying a diverse set of problems, including cue combination, decision-making, and visual attention.

10.2 Introduction

A central question in computational neuroscience concerns the information contained in the response of a population of neurons. In the highly constrained and controlled situations typically studied in neurophysiological experiments, the relevant information can often be described as a single underlying stimulus variable. For example, when a monkey is faced with a series of visual stimuli, each of which contains a set of randomly placed dots all moving in a particular direction, the relevant information in a population of cells that respond selectively to motion direction is that single, unambiguous direction of motion.

In a more natural situation, however, the responses of a population of neurons are influenced by a variety of stimulus aspects, and the information about any particular variable is unlikely to be uniquely specified, with total certainty.

The aim of this chapter is to show that one needs to adopt a different view of the information contained in population in order to properly understand the neural code. The representational capacity of a population must be extended, as including not just a single value, but instead preserving information about all possible values, and the degree to which each is consistent with the responses. The key hypothesis, then, that drives this approach is that neural populations represent and in effect compute probabilities.

10.3   An Example

Many investigators have examined neural and behavioral responses to stimuli composed of two patterns sliding across each other. These often create the impression of two separate surfaces moving in different directions. Neurophysiologists have focused on the responses of cells in the medial temporal (MT) area of monkeys to stimuli such as these, since these cells are selective for particular directions of motion, and are well activated by such random displays provided that some of the dots are moving in the directions that they prefer.

The general neurophysiological finding is that an MT cell's response to these stimuli can be characterized as the average of its responses to the individual components (van Wezel et al., 1996; Recanzone et al., 1997). As an example, Figure 10.1 shows data obtained from single-cell recordings in MT to random dot patterns consisting of two distinct motion directions (Treue et al., 2000). Each plot is for a different relative angle ($\Delta\theta$) between the two directions. A plot can equivalently be viewed as the response of a population of MT cells having different preferred directions to a single presentation of a stimulus containing two directions. If $\Delta\theta$ is large, the activity profile is bimodal, but as the directional difference shrinks, the profile becomes unimodal. The population response to a $\Delta\theta = 30°$ motion stimulus is merely a wider version of the response to a stimulus containing a single direction of motion. However, this transition from a bimodal to unimodal profiles in MT does not apparently correspond to subjects' percepts; subjects can reliably perceive both motions in superimposed transparent random patterns down to an angle of $10°$ (Mather and Moulden, 1983). If these MT activities play a determining role in motion perception, the challenge is to understand how the visual system can extract both motions from such unimodal (and bimodal) response profiles.

Unlike the situation when the stimulus is a set of random dots all moving coherently in the same direction, so that the relevant information is that single direction of motion, in the transparency case the relevant information is two directions of motion.

This seemingly simple change turns out to be crucial for methods of extracting information, or decoding the population response. As we show below, these methods fail
to recover both motions from population responses such as those shown in Figure
10.1, essentially because they contain an inherent assumption that a single value
underlies the response. In addition, these methods fail to recover the relevant information in other random-dot motion stimuli slightly more complicated than the coherent uni-directional stimulus, such as when the dot directions are chosen from a
Gaussian distribution about a particular dominant direction.

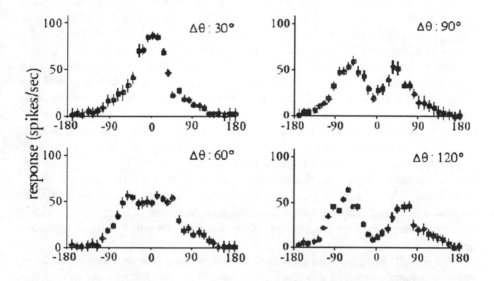

Figure 10.1. Each of the four plots depicts a single MT cell response (spikes per second) to a
transparent motion stimulus of a fixed directional difference ($\Delta\theta$) between the two motion
directions. The x-axis gives the average direction of stimulus motion relative to the cell's preferred direction (0°). Reproduced by permission from Treue et al. (2000). © Nature Publishing
Group, 2000.

The random dot motion stimuli are intended as simple examples that have been
explored in neurophysiological studies. The idea that neural populations are best
viewed as encoding probabilities over variables rather than single unambiguous values is relevant not just in these carefully constructed stimulus situations, but instead
applies widely to many natural situations, particularly when stimuli become less constrained and more complex.

For example, uncertainty must be taken into account when one considers the general case of image motion, and velocity encoding in local regions. The response of a
population of MT cells, when only a single straight edge is moving in their common
receptive field, contains an inherent ambiguity – several two-dimensional velocities
are consistent with the velocity of any moving one-dimensional pattern (e.g., a long
edge or grating). This ambiguity is known as the "aperture problem", since the

motion of a one-dimensional stimulus viewed through a small circular aperture is ambiguous. In such cases where the velocity is inherently ambiguous, extracting a single velocity from the MT population code is impossible. In locations where the stimulus contains more spatial structure, the motion may be disambiguated. However, when noise is considered, all locations will have some degree of ambiguity. The system then needs to quantify the degree to which the data is consistent with a particular velocity. In addition, uncertainty about local velocity stems from sources other than local geometry, including contrast and duration. At high contrast, the image information may unambiguously support the true velocity. But at lower contrasts, many velocities may be consistent with the image information, and distinguishing the true velocity from others will be increasingly difficult as contrast is reduced. Similarly, at short durations distinguishing the true velocity will also be difficult, and a range of velocities will be consistent with the image information.

The second aspect of image information that motivates the interpretation of a population response as a full distribution is the presence of multiple values within a receptive field. Transparency is just one example of this. Another example concerns higher levels of the visual system. For example, in the inferotemporal (IT) cortex, cells have large receptive fields, and in a natural scene containing multiple objects, more than one object is likely to lie within the shared receptive fields of a single IT population. Any description of the objects that can account for that population response would thus need to include support for each of the objects present. In this case, as with the case of uncertainty, a more natural description of the information contained in the population response is not a single best value of the relevant variable but rather the probability that various values are each consistent with the observed response.

The rest of this chapter is organized as follows: we first describe methods by which a population of cells can represent a probability distribution over an underlying value. We then show how this approach is useful for elucidating cortical function in a number of different ways, including analyzing responses of a single cortical population, understanding lateral interactions within a population, and characterizing information processing from one population to another.

10.4   Representing Distributions in Populations

From the standpoint of information processing, representing probabilities rather than just single values can lead to significant improvements. Consider an abstract view of cortex as a multiple-level processor where information converges and diverges from several populations at one level to several at the next. If at one level the information in the population response is reduced to a single value, then downstream populations will not have access to the uncertainty of the estimate, and to the likelihood of other values. This uncertainty may be important, particularly as information from other populations is included, to ensure that potentially better values are not ignored. Representing the degree of support across the range of values prevents premature commitment, and enables better judgements to be made downstream.

A key point with respect to cortical representations of probabilities is that the probabilities need not be explicitly decoded from a population response. If the probabilistic information contained in one population can give rise to probabilistic information in another population, no explicit decoding ever needs to take place. Thus the focus in this chapter is not on neural mechanisms of extracting distributions from a population response (i.e., how subsequent stages of neural processing might actually extract information from the population), but rather on methods of establishing in a more abstract way the information that can be shown by any means to be present (a similar approach is pursued for decoding in individual neurons in Rieke et al., 1997).

Methods for representing a probability distribution in a population can be divided into two classes: basis functions, and generative methods. Before describing these two classes, we introduce some notation. Associated with each cell in the population is a *tuning function* $f_i(x)$, which describes the expected response of that cell as a function of the value of the relevant variable. These tuning functions are typically determined using simple stimuli (e.g., ones containing uni-directional motion). An important point is that the methods discussed here make no assumptions about the form of these tuning functions; rather, all that they require is some description of how the cell response varies as a function of the variable value. The aim then is to form a distribution $\hat{P}^r(x)$ over these variable values based on the response vector r. The elements in the response vector, r_i could, for example, be the number of spikes that cell i emits in a fixed time interval following the stimulus onset. Equivalently, r_i could be considered the rate parameter of a Poisson distribution that describes the neuron's probability of emitting a spike. The goal is to decode the collection over values that are consistent with these responses on that trial.

10.5   Basis Function Representations

Simoncelli and Heeger (1994, 1998) proposed a model in which MT cell responses were viewed as spatiotemporal filters that could be used to derive not a single local estimation of motion but instead a local likelihood map, i.e., a distribution over velocities. Weiss and Adelson (1998) have recently adopted a similar approach to model a wide range of results in visual motion psychophysics. Basis function methods have also been proposed as a way of understanding population responses in parietal cortex (Pouget and Sejnowski, 1997). A general theory of using basis functions as a means to interpret population codes as representing probability distributions was developed by Anderson (1994) and Anderson and Van Essen (1995).

A basis function approach represents the distribution $\hat{P}^r(x)$ as a linear combination of simple kernel or basis functions $\psi_i(x)$ associated with each cell, weighted by a normalized function of its activity r_i:

$$\hat{P}^r(x) = \sum_i r'_i \, \psi_i(x).\tag{1}$$

Here the r'_i are normalized such that $\hat{P}^r(x)$ is a probability distribution. If the $\psi_i(x)$ are probability distributions themselves, and r_i are all positive, a natural choice is to have

$$r'_i = \frac{r_i}{\sum_j r_j}. \tag{2}$$

Note that the kernel functions $\psi_i(x)$ are *not* the tuning functions $f_i(x)$ of the cells that would commonly be measured in an experiment. They need have *no* neural instantiation; instead, they form part of the interpretive structure for the population code.

10.6   Generative Representations

The basis function approach can be viewed as approximating a probability distribution by adding together distributions associated with each neuron. The second class of models also builds a distribution by adding pieces together, but the addition is done in the domain of log probabilities (G. Hinton, personal communication). This overcomes an important limitation of the basis function approach, which cannot represent distributions that are sharper than the component distributions.

This second class of models adopts a *generative* approach, where the goal is to find a value or set of values of the underlying variable that could have generated the observed population response r. These methods begin by hypothesizing an *encoding* model, that describes how it is encoded in the noisy elements of the population response. This encoding model specifies $P[r|x]$, a distribution over response patterns given the variable. They then apply a *decoding* algorithm to estimate the distribution over x, typically using Bayes' Rule to invert the relationship between r and x.

10.7   Standard Bayesian Approach

A standard approach to decoding population responses turns out to be a type of generative representation. This approach has enjoyed some success in applications to decoding actual population responses. For example, Zhang et al. (1998) applied the Bayesian method to estimate the position of a rat on a maze based on spike counts from approximately 30 hippocampal place cells; the estimation accuracy was within the accuracy range of the position sensor on the rat.

Standard Bayesian decoding starts with the activities $r = \{r_i\}$ and generates a distribution $P[x|r]$. The simplest model assumes that the response of each cell i in the population is Poisson where $f_i(x)$ is the mean rate, and that the noise in these responses is independent. Under this model:

$$L\left[\hat{P}^r(x)\right]= \log P[x\mid r] \sim \log\left\{P[x]\prod_i P[r_i\mid x]\right\} \sim \sum_i r_i \log f_i(x). \quad (3)$$

This method is also a form of basis function estimate, but it is *multiplicative*, because it is additive in the log domain. It thus tends to produce a sharp distribution for a single value of x. In the standard approach, a single estimate is extracted from $P[x\mid r]$. However, this final step is not necessary, and the information about all possible values of x can be maintained.

By comparing Equations 1 and 3, one sees that the decoding in both cases involves a weighted sum of each cell's tuning function (or log thereof), weighted by its observed spike count. In the generative case, however, this quantity is proportional to the log of the distribution, which allows the resulting distribution to be concentrated on specific values of the underlying variable.

10.8 Distributional Population Coding

The standard Bayesian encoding model assumes that the population responses are caused by a single value x^* of the relevant underlying variable: $\langle r_i\rangle = f_i(x^*)$. Under this model, Bayesian decoding takes the observed activities and produces probability distributions over x (Equation 3).

Our extended Bayesian distributional population coding (DPC) method (Zemel et al., 1998) extends the standard encoding model to allow r to depend on general $P[x]$:

$$\langle r_i\rangle = \int_x P[x]f_i(x)dx. \quad (4)$$

Under this model, Bayesian decoding takes the observed activities r and produces probability distributions over probability distributions over x, $P[P(x)\mid r]$. For simplicity, we decode using an approximate form of maximum likelihood in distributions over x, finding the $\hat{P}^r(x)$ that maximizes

$$L\left[\hat{P}^r(x)\right] \sim \sum_i r_i \log\left[f_i(x) * \hat{P}^r(x)\right] - \alpha g\left[\hat{P}^r(x)\right] \quad (5)$$

where the smoothness term $g[\ ]$ acts as a regularizer.

The distributional encoding operation in Equation 4 is quite straightforward – by design, since this represents an assumption about what neural processing prior to the population under study performs. However, the distributional decoding operation that we have used (Zemel et al., 1998) involves complicated and non-neural operations. The idea is to understand what information in principle may be conveyed by a population code under this interpretation, and then to judge actual neural operations in the light of this theoretical optimum. Our extended Bayesian approach is a statistical cousin of so-called line-element models, which attempt to account for subjects' performance in cases like transparency using the output of some fixed number of direction-selective mechanisms (Williams et al., 1991).

Some intuition into the relationship between the two generative models can be gained by inspecting Equations 3 and 5. Both models can be viewed as forming $\hat{r}$, a predicted value of r based on some function of x, to match the actual observed rates r: the likelihood

$$L = \sum_i r_i \log \hat{r}_i$$

is maximized when $\{\hat{r}_i\}$ matches $\{r_i\}$. In the standard method, a single value of x gives rise to the observed rates ($\hat{r}$ is based on a single value of x), but in the extended method, a full distribution over x gives rise to the rates, as $f_i(x)$ is convolved with the current estimate $\hat{P}^r(x)$. This latter approach fits with the objective of finding the best collection over values of x to match the actual population response.

10.9 Applying Distributional Population Coding

10.9.1 Population Analysis

We have applied these methods in a number of domains. The most straightforward application of these methods is to decode distributions from the response of a population of neurons. Below we briefly describe two such applications.

10.9.2 Decoding Transparent Motion

We have applied these methods to simulated MT population responses to multiple motion stimuli, such as the responses shown in Figure 10.1. We summarize the results here; further details may be found in Zemel and Dayan (1999). The underlying variable of interest in this domain is the motion direction(s), so $x = \theta$. In the simulations described here, we have used a population of 200 model MT cells, with tuning functions defined by random sampling within physiologically determined ranges for the parameters: baseline b, amplitude a, and width σ. The encoding model comes from the MT data: for a single motion

$$\langle r_i | \theta \rangle = f_i(\theta) = b_i + a_i \times \exp\left[-\frac{(\theta - \theta_i)^2}{2\sigma_i^2} \right],$$

while for two motions,

$$\langle r_i | \theta_1, \theta_2 \rangle = \frac{1}{2}[f_i(\theta_1) + f_i(\theta_2)].$$

The noise is taken to be independent and Poisson. For multiple motion stimuli, this encoding model produces the observed neurophysiological response: each unit's expected activity is the average of its responses to the component motions.

For the distributional population coding methods to decode successfully when there are two motions in the input (θ_1 and θ_2) the extracted distribution must at least have two modes. Standard Bayesian decoding (Equation 3) fails to satisfy this requirement. First, if the response profile r is unimodal (Figure 10.1, the 30° plot), convolution with unimodal kernels $\{\log f_i(\theta)\}$ produces a unimodal $\log P[\theta | r]$, peaked about the average of the two directions. The basis function method (Equation 1) suffers from the same problem, and also fails to be adequately sharp for single value inputs.

Surprisingly, the standard Bayesian decoding method also fails on bimodal response profiles. If the baseline response $b_i = 0$, then $P[\theta | r]$ is Gaussian, with mean

$$\frac{\sum_i r_i \theta_i}{\sum_i r_i}$$

and variance

$$\frac{1}{\sum_i r_i / \sigma_i^2}$$

(Snippe, 1996; Zemel et al., 1998). If $b_i \succ 0$, then, for the extracted distribution to have two modes in the appropriate positions, $\log[P[\theta_1 | r] / P[\theta_2 | r]]$ must be small. However, the variance of this quantity is

$$\sum_i \langle r_i \rangle (\log [f_i(\theta_1) / f_i(\theta_2)])^2,$$

which is much greater than zero unless the tuning curves are so flat as to be able to convey only little information about the stimuli. Intuitively, the noise in the rates causes

$$\sum r_i \log f_i(\theta)$$

to be greater around one of the two values, and exponentiating to form $P[\theta | r]$ selects out this one value. Thus the standard method can only extract one of the two motion components from the population responses to transparent motion.

Our extended Bayesian method, on the other hand, matches the generating distribution for bimodal response patterns (Figure 10.2). For unimodal response patterns, such as those generated by double motion stimuli with $\Delta\theta = 30°$, DPC also consistently recovers the generating distribution. The bimodality of the reconstructed distribution begins to break down around $\Delta\theta = 10°$, which is also the point at which subjects are unable to distinguish two motions from a single broader band of motion directions (Mather and Moulden, 1983).

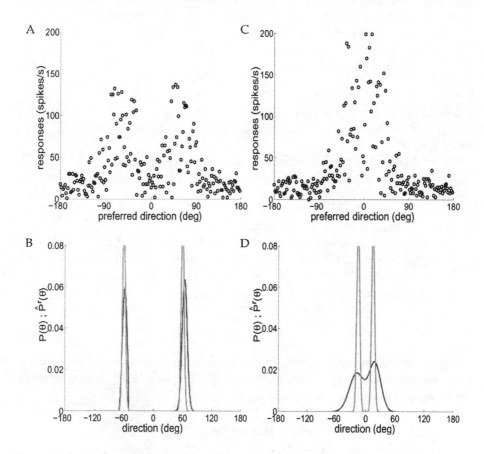

Figure 10.2. (A) On a single simulated trial, the population response forms a bimodal activity profile when $\Delta\theta = 120°$; (B) the reconstructed (darker) distribution closely matches the true input distribution for this trial; (C) as $\Delta\theta \rightarrow 10°$, the population response is no longer bimodel, instead has a noisy unimodal profile; and (D) the reconstructed distribution no longer has two clear modes.

It has been reported (Treue et al., 2000) that for angles $\Delta\theta < 10°$, subjects can tell that all points are not moving in parallel, but are uncertain whether they are moving in two discrete directions or within a directional band. Our model qualitatively captures this uncertainty, reconstructing a broad distribution with two small peaks for directional differences between 7° and 10°. Our model also matches psychophysical performance on other aspects of these stimuli, and generates a number of testable predictions.

10.9.3   Decision Noise

A second application of distribution population coding concerns a well-studied set of psychophysical and neurophysiological results for a dot-motion task in which monkeys must discriminate the single coherent direction of motion amidst a random motion background. These data pose a notorious puzzle for population coding: it seems that there are *single* cells in monkey MT that are just as accurate as the whole monkey, i.e., an ideal observer judging the direction of stimulus motion based on the response of optimally stimulated MT neurons could match the monkey's psychophysical performance (Britten et al., 1992). The puzzle is that monkeys are not much better than their single cells, since one would expect them to be able to average away substantial noise. That is, neural coding seems unreasonably inefficient. Shadlen et al. (1996) analyzed this puzzling inefficiency and proposed a model, which includes a number of factors that contribute to the poor population performance. Even after taking account of factors such as correlations between the activities of cells with similar orientation preferences, their model requires the addition of extra noise in the decision-making stage to reduce the accuracy appropriately. Distributional population coding offers an alternative hypothesis to account for the apparent inefficiency: that the activity of the population should be seen as coding the whole distribution of inputs rather than just a single direction of motion in the face of noise. Here we outline the application of this approach to the decision-noise paradigm.

The experimental paradigm involves measuring neuronal responses while monkeys perform a two-alternative forced discrimination of motion direction. The monkeys choose from opposite directions in displays containing a unidirectional motion signal embedded in a field of random motion noise. The axis of the direction discrimination is tailored to the preferred direction of each individual neuron (but we will call it "up" or "down"). The *coherence*, or strength of the signal is varied by changing the percentage of dots moving in the single direction.

In the Shadlen et al. (1996) model, the final decision about direction is based on pooling the activities across the population (Figure 10.3A). The details of the model include correlated noise in the activity of MT neurons; scaled inclusion in the pool of neurons whose preferred direction did not match the discrimination direction; and, crucially, noise in the output of each pool. The final decision was based on which pool had the larger value (Figure 10.3A).

We attribute the apparent inefficiency to a different cause, suggesting that it is unreasonable to think of the output of all the neurons as being combined directly to make a decision about one of two directions of motion. Rather, the population can be taken as encoding extra information, namely a whole distribution over the directions of motion in the input, and inferences about whether there is more "up" than "down" in the input have to be based on the distribution that can be decoded from this (see Figure 10.3B). To put it another way, the population only looks inefficient if it is treated as coding the proportions of motion in two directions, not if it is treated as coding the proportions of motion in all directions.

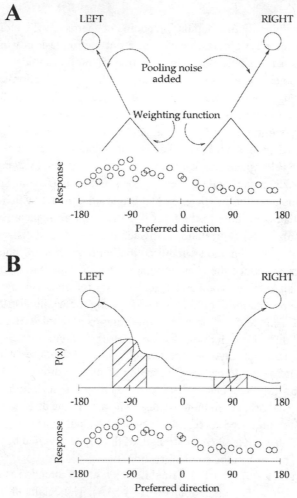

Figure 10.3. (A) The value assigned to each pool in the Shadlen et al. (1996) model is a random variable whose mean is a weighted combination of responses of neurons whose preferred directions are close to the target direction. The larger pooled signal dictates the response. (B) The output of the extended Bayesian DPC model is the integral under the inferred distribution within a window around the target direction.

We have applied the extended Bayesian distributional population coding (DPC) model to this domain. Figure 10.4A shows that the model's performance degrades as the stimulus coherence decreases, in a manner consistent with the monkey's psychophysical performance. Figure 10.4B shows how the model's performance depends on the radius of the window around the target direction, as well as the number of units in the population. This plot suggests a bias/variance trade-off in the size of the window – a larger window involves more units in the target discrimination (through the decoding process), but also increases the effect of the motion noise.

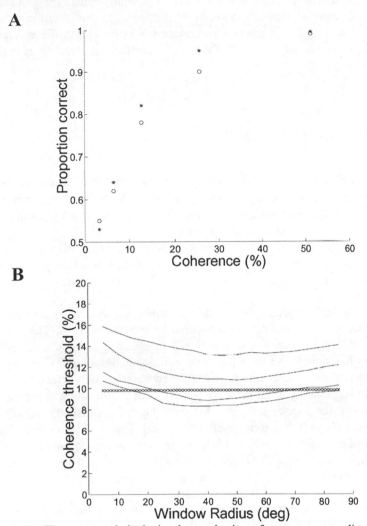

Figure 10.4. (A) The open symbols depict the monkey's performance on one direction discrimination experiment as a function of stimulus strength (Britten et al., 1992). The closed symbols show the proportion correct of the DPC model; each symbol is the average of 200 trials in which the unit responses were determined by the encoding model (Equation 4). The integration window radius is 20°, and population contains 100 units. Britten et al. summarize these results by a fitting function of the form $p = 1 - 0.5 \exp[-(c/\alpha)^\beta]$ to the points where p is the proportion correct, c is the coherence level, α is the coherence level supporting "threshold" performance (82% correct), and β is the function slope. The key empirical parameter that Shadlen et al. (1996) attempt to model is α. (B) The output of the DPC model depends on the radius of the integration window around each target direction and the number of units in the populatiion ($N = 50, 100, 150, 200$ from top to bottom). The model's performance is summarized by the threshold coherence level (α). The horizontal line is the empirical mean $\alpha = 9.8\%$ reported in Shadlen et al. (1996).

These results demonstrate that our model is generally consistent with the monkey performance over a range of parameter values; yet, like the penultimate models in Shadlen et al. (1996), for the best parameter values it exceeds the accuracy of the monkeys. However, we have not yet included the effects of correlations in the activities of cells.

10.9.4   Lateral Interactions

Distributional population coding can provide a method of analyzing population responses in domains other than motion processing. We briefly summarize one application here, concerning another well-studied area in population coding: orientation tuning in primary visual cortex. One class of network models posits that the sharp tuning of striate cells arises from broadly tuned feed forward LGN inputs which are substantially sharpened by *recurrent* connections within the striate population. These *recurrent* models can account for a wide range of the empirical data (reviewed in Sompolinsky and Shapley, 1997).

Carandini and Ringach (1997) studied a simplified recurrent model that replicates many observed characteristics of a hypercolumn of striate cells. The core of their model is a single equation implementing a feedback filter with a center-surround weighting function. When the input to the cell population contains two orientations, their model exhibits peculiar responses: if the two orientations differ by less than 45°, the model responds as if the input contained a single orientation at the mean of the two values; if the angular difference between the two orientations exceeds 45°, the model responds as if they were nearly orthogonal. The model also cannot signal the presence of three orientations, and creates spurious orthogonal orientations to noisy single orientation inputs. These authors analyzed their model and showed tha these effects of attraction and repulsion between orientations are unavoidable in a broad class of recurrent models of orientation selectivity.

This model thus makes strong predictions about the visual response of cortical neurons to stimuli containing multiple orientations. Relevant empirical data are scarce. DeAngelis et al. (1992) found that the response of a striate cell to its optimal orientation is reduced when the stimulus contains a second orientation. Cell responses to pairs of non-optimally oriented stimuli are not known. Nonetheless, the predictions of the model appear dubious: it predicts that the neurons will be unable to veridically encode multiple orientations, even when the orientations differ by nearly 45°.

We have proposed an alternative model (Zemel and Pillow, 1999) that is able to maintain a broader range of activity patterns in the population for multiple orientation stimuli. The dynamics in this model are very similar to the Carandini-Ringach model, with three modifications. Our model utilizes optimized recurrent weights in the population; shunting inhibition of the excitatory potential; and a nonlinear activation function.

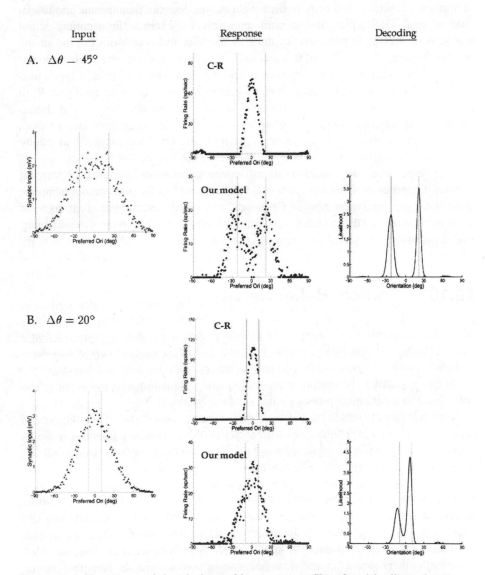

Figure 10.5. The top row of plots depicts stable response profiles of model cells to stimuli containing two orientations. The three columns correspond to differing angular spread between the pair of orientations. In the top row of plots, the thicker dots show the cell responses in our model, the thinner dots show the responses of the Carandini-Ringach model to the same inputs. The bottom row of plots depicts the information about orientation encoded from those response profiles. Here the thin dotted lines are the true orientations in the model inputs, and the thick lines are the orientations encoded in the model responses. The contrast between the models' responses in the middle column shows that ours can stably maintain bimodal activity patterns even when the modes are not orthogonal. And information about both orientations is retained even as the angular difference shrinks below $45°$.

The model uses the extended Bayesian DPC method to determine what information about orientation is present in the model cell responses. This approach is essential in this case, since the central question is whether information about multipleorientations is preserved in the resulting patterns. As was the case in the motion domain, a key feature of this method is that unimodal response profiles do not necessarily indicate that only one value is present in the input; instead, a broad unimodal profile is consistent with a stimulus containing two nearby orientations. With respect to DPC, a novel aspect of this application is that the encoding model (Equation 4) is used to optimize the recurrent weights within the population: target values are obtained by convolving the tuning functions (preferred orientation and tuning width) of the model striate cells with the set of orientations present in the stimulus.

The response of the model to stimuli containing two orientations, of varying angular difference, is shown in Figure 10.5. The Carandini-Ringach model is unable to veridically encode these stimuli. Our model forms different stable activity patterns to all three pairs, and the decoding method forms a bimodal distribution in each case, suggesting that two orientations are present in the stimulus.

10.10   Cortical Belief Network

DPC provides a method of interpreting general probability distribution from a population of units representing a particular variable. The larger goal is to understand how probabilities, or distributional information in one population can affect the responses of another population. A network composed of many populations representing different, related variables must properly combine distributions.

This problem of combining distributions can be cast as a formulation of a type of belief network, in which associated with each population code is a probability distribution over a particular variable or set of variables, and the links between the populations express conditional dependencies between these variables.

Consider a simple situation in which an animal is stationary while an object is moving and the animal wants to estimate time-to-collision. Various forms of information represented in different neural populations are useful in this task, including one population of cells representing the direction of motion of an object, and a second population encoding the distance from the object to the animal. These two variables, along with others such as the object speed, can be used to infer the time-to-collision of the object, also encoded in a population of neurons.

We cast this problem in the probabilistic framework of a belief network, which sets up a description of the computation to be implemented in populations of neurons. The cortical belief network framework for this example is cartooned in Figure 10.6. Links between a population encoding direction and a second population encoding time-to-collision (T) are the typical relations between these variables. For example, a very short T value is consistent only with object motions directly towards the animal, while longer times are consistent with a range of directions. Based on these conditional relations, and assuming that direction and distance are conditionally independent given time-to-collision, then a distribution over T can be inferred from distributions over the other two variables.

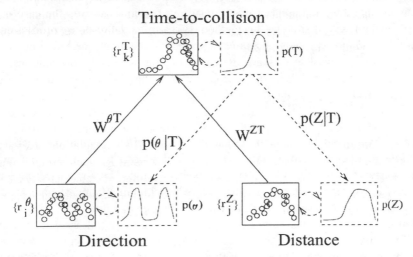

Figure 10.6. An example of a cortical belief network formulated to combine population code representations of probability distributions. The circular elements in the solid boxes depict explicit network components (neurons), while the curves in the dashed boxes depict implicit interpretations of those neural responses.

Given true probability distributions $P[\theta|\omega]$ and $P[Z|\omega]$ over the direction (θ) and distance (Z) of the object (here ω represents the underlying information in the environment, including visual and extra-retinal cues), inferring time-to-collision (T) requires calculating

$$P[T|\omega] = \int_{\theta,Z} P[\theta|\omega] P[Z|\omega] P[T|\theta,Z] \, d\theta dZ$$
$$\alpha P[T] \int_{\theta,Z} P[\theta|\omega] P[Z|\varpi] P[\theta,Z|T] \, d\theta dZ \qquad (6)$$

where $P[T]$ is the prior distribution over T. This equation governs computation in the belief network (shown in dashed lines in Figure 10.6), which forms a target for the computation carried out in the neurons.

We have therefore reduced the computational problem to one of mapping activities r^θ and r^Z into activities r^T for which $P[T|r^T]$ is a good approximation to the integration in Equation 6, where $P[\theta|\varpi]$ is what r^θ represents and $P[Z|\varpi]$ is what r^Z represents.

An important point is that the belief network – the probability distributions and the conditional links – are not actually implemented. Instead it provides an interpretation for the network of neurons. Thus this framework provides a dual representation in which the probabilities are implicitly encoded, while the population responses are explicit.

This model makes explicit predictions as to how information from different sources are combined as the values and certainty of the sources are manipulated. We are currently testing these predictions in psychophysical experiments on curvature judgments based on a combination of shading and texture cues. Preliminary results suggest that the model provides a natural explanation for a diverse set of phenomena, including cue enhancement and veto.

10.11   Discussion

The main hypothesis advanced in this chapter is that cortical populations encode distributions or likelihoods over variables. Standard approaches consider that, at any point in time, a population of neurons represents a single "best-guess" about a relevant variable or set of variables. An alternative novel interpretation of cortical activities is that they jointly specify likelihood across a range of values.

Different forms of such representations have been proposed, each with a corresponding method of extracting probability distributions from neural activities. A basis function approach (Anderson, 1995), and a standard Bayesian model are both capable of decoding distributions, but neither approach explicitly contains a description of how the population encodes uncertainty or multiple values. We have proposed a model, distributional population coding (DPC), that does contain an explicit decoding model, and formulates the decoding problem as the Bayesian inverse of this encoding model.

We then examined the performance of these various representational schemes to an example problem. Transparent motion provides an ideal paradigm for population coding of probability distributions: multiple values are encoded in each trial, making standard single-value interpretation inapplicable, and the combination of physiological data and behavioral data respectively specify neural activities and information presumably inferred from those activities. The basis function model and the standard Bayesian model both perform poorly in this paradigm. DPC, which treats neuronal responses and the animal's judgments as being sensitive to the entire distribution of an encoded variable, has been shown to be consistent with both single-cell responses and behavioral decisions, even matching subjects' threshold behavior. DPC also provides a novel perspective on another motion paradigm, suggesting that the judgement as to the single coherent motion direction amidst random motion is not hampered by decision noise but instead sensitive to the full array of motion directions present during the experimental trial.

We are currently applying this same model to several other motion experiments, including one in which subjects had to determine whether a motion stimulus consisted of a number of discrete directions or a uniform distribution (Williams et al., 1991). We are investigating whether our model can explain the non-monotonic relationship between the number of directions and the judgments. Finally, experiments showing the effect of target uncertainty on population responses (Basso and Wurtz, 1998; Bastian et al., 1998) are also handled naturally by the DPC approach.

DPC also provides a new approach to the issue of orientation tuning in the primary visual cortex. It suggests that the representational capability of a V1 population should be expanded to include multiple values and uncertainty. A consequence of this view are specific predictions as to how a V1 population will respond to images containing ambiguous (low-contrast, noisy, and/or curved) edges, or containing multiple edges. We have developed a recurrent model of orientation tuning within the framework of the DPC approach, which is capable of encoding a range of stable activity profiles. This new model makes a number of testable predictions concerning V1 responses that differ from another recent model.

Finally, the current model is intended to describe the information available in a population at one stage in the processing stream. Extending the model to address how this information gets into the population, and the neural decoding and decision mechanisms subsequent to it, requires constructing a network of populations, each encoding distributions over different variables. A cortical belief network, introduced in the final section, is a method for carrying out probabilistic inference in populations of neurons.

10.12   Acknowledgements

Thanks to Peter Dayan, Geoffrey Hinton, Jonathon Pillow, Dave Towers, and Zhiyong Yang for useful discussions, and to Stefan Treue for providing the data plot and for informative discussions of his experiments. Supported by ONR N00014-98-1-0509.

References

Anderson, C.H. (1994) Basic elements of biological computational systems. International Journal of Modern Physics C 5(2): 135–137.

Anderson, C.H. (1995) Unifying perspectives on neuronal codes and processing. In: XIX International workshop on condensed matter theories. Caracas, Venezuela.

Anderson, C.H., Van Essen, D.C. (1995) Neurobiological computational systems. In: IEEE World Congress on Computational Intelligence, pp. 213-222.

Basso, M.A., Wurtz, R.H. (1998) Modulation of neuronal activity in superior colliculus by changes in target probability. Journal of Neuroscience 18(18): 7519–7534.

Bastian, A., Riehle, A., Erlhagen, W., Schoner, G. (1998) Prior information preshapes the population representation of movement direction in motor cortex. Neuroreport 9(2): 315–319.

Britten, K.H., Shadlen, M.N., Newsome, W.T., Movshon, J.A. (1992) The analysis of visual motion: A comparison of neuronal and psychophysical performance. Journal of Neuroscience 12(12): 4745–4765.

Carandini, M., Ringach, D. (1997) Predictions of a recurrent model of orientation selectivity. Vision Res 37: 3061–3071.

DeAngelis, G.C., Robson, J.G., Ohzawa, I., Freeman, R.D. (1992) Organization of suppression in receptive fields of neurons in cat visual cortex. J. Neurophysiol. 68: 144–163.

Grunewald, A. (1996) A model of transparent motion and non-transparent motion aftereffects. In: D.S. Touretzky, M.C. Mozer, M.E. Hasselmo (Eds.) Advances in Neural Information Processing Systems 8. Cambridge, MA: MIT Press, pp. 837–843.

Hol, K., Treue, S. (1997) Direction-selective responses in the superior temporal sulcus to transparent patterns moving at acute angles. Society for Neuroscience Abstracts 23: 459.

Mather, G., Moulden, B. (1983) Thresholds for movement direction: two directions are less detectable than one. Quarterly Journal of Experimental Psychology 35: 513–518.

Pouget, A., Sejnowski, T.J. (1997) Spatial transformations in the parietal cortex using basis functions. Journal of Cognitive Neuroscience 9(2): 222–237.

Rauber, H.J., Treue, S. (1997) Recovering the directions of visual motion in transparent patterns. Society for Neuroscience Abstracts 23: 179.

Recanzone, G.H., Wurtz, R.H., Schwarz, U. (1997) Responses of MT and MST neurons to one and two moving objects in the receptive field. Journal of Neurophysiology 78(6): 2904–2915.

Rieke, F., Warland, D., de Ruyter van Steveninck, R.R., Bialek, W. (1997) Spikes: Exploring the Neural Code. Cambridge, MA: MIT Press.

Shadlen, M.N., Britten, K.H., Newsome, W.T., Movshon, J.A. (1996) A computational analysis of the relationship between neuronal and behavioral responses to visual motion. Journal of Neuroscience 16(4): 1486–1510.

Simoncelli, E., Adelson, E., Heeger, D. (1991) Probability distributions of optical flow. In: Proceedings 1991 IEEE Computer Society Conference on Computer Vision and Pattern Recognition, pp. 310–315.

Simoncelli, E., Heeger, D. (1994) A velocity representation model for MT cells. Investigative Opthamology and Visual Science Supplement 35: 1827.

Simoncelli, E., Heeger, D. (1998) A model of neuronal responses in visual area MT. Vision Research 38(5): 743-761.

Snippe, H.P. (1996) Theoretical considerations for the analysis of population coding in motor cortex. Neural Computation 8(3): 29–37.

Sompolinsky, H., Shapley, R. (1997) New perspectives on the mechanisms for orientation selectivity. Curr. Opin. Neurobiol. 7: 514–522.

Treue, S., Hol, K., Rauber, H.-J. (2000) Seeing multiple directions of motion: Physiology and psychophysics. Nature Neuroscience 3(3): 271–277.

van Wezel, R.J., Lankheet, M.J., Verstraten, F.A., Maree, A.F., van de Grind, W.A. (1996) Responses of complex cells in arca 17 of the cat to bi-vectorial transparent motion. Vision Research 36(18): 2805- 2813.

Weiss, Y., Adelson, T. (1998) Slow and smooth: Combination of local motion signals. Vision Research 31(2): 275–286.

Williams, D., Sekuler, R. (1984) Coherent global motion percepts from stochastic local motions. Vision Research 24(1): 55–62.

Williams, D., Tweten, S., Sekuler, R. (1991) Using metamers to explore motion perception. Vision Research 31(2): 275–286.

Zemel, R.S., Dayan, P., Pouget, A. (1998) Probabilistic interpretation of population codes. Neural Computation 10: 403– 430.

Zemel, R.S., Dayan, P. (1999) Distributional population codes and multiple motion models. In: M.S. Kearns, S.A. Solla, D.A. Cohn (Eds.) Advances in Neural Information Processing Systems 11. Cambridge, MA: MIT Press.

Zemel, R.S., Pillow, J. (1999) Encoding multiple orientations in a recurrent network. Proceedings of the Computational Neuroscience Society, 8, Kluwer Press.

Zhang, K., Ginzburg, I., McNaughton, B.L., Sejnowski, T.J. (1998) Interpreting neuronal population activity by reconstruction: A unified framework with application to hippocampal place cells. Journal of Neurophysiology 79(2): 1017–1044.

Index

ACTION network   232, 236-7, 239,
   247-8, 250-4, 256, 261
Action potential (AP)   85-6, 126-
   8,130, 133
ACT-R *see* cognitive architecture
AI *see* artificial intelligence
ANN *see* artificial neural network
AM *see* amplitude modulation
ANS *see* autonomic nervous system
AP *see* action potential
ARAS *see* ascending reticular activat-
   ing system
ART *see* adaptive resonance theory
Absence *see* epilepsy
Action
   generation   100
   prediction   2-3, 11, 14-21
   representation module   255
   potential (AP)   136-7, 140-4,148-
         50, 15-6, 160-4
   schemas   247
   sequences   254
Activation vigor   110
Adaptive control of thought-rational
   (ACT-R) *see* cognitive architec-
   tures
Adaptive resonance theory (ART)   101
Adaptive synapses   223

Algorithm   191, 194, 224, 236, 240,
   272
   classifier   190
   hierarchical clustering   190
   Leabra   194
   learning   192
Amnesia *see* epilepsy
Amplitude modulation (AM)   71-74
Amygdala   80, 185, 232
Anatomical
   basis   68, 82
   evidence   79
   properties   206
   structure   145
Antecedent support network   94
Anterior nucleus   71
Aperture problem   269
Architectural principles   212
Artificial intelligence (AI)   58, 65-6,
   91, 171, 175, 184
   nervous system   22
   neural network (ANN)   29, 36, 66,
         174, 178, 184, 186, 212
   sensing system   196
Ascending reticular activating system
   (ARAS)   236
Associationism   56-59